OUT THERE LEARNI.

Critical Reflections on Off-Campus Study Programs

Universities across North America and beyond are experiencing growing demand for off-campus experiential learning. Exploring the foundations of what it means to learn "out there," *Out There Learning* is an informed critical investigation of the pedagogical philosophies and practices involved in short-term off-campus programs or field courses. Bringing together contributors' individual research and experience teaching or administering these programs, *Out There Learning* examines and challenges common assumptions about pedagogy, place, and personal transformation, while also providing experience-based insights and advice for getting the most out of faculty-led field courses.

Divided into three sections that investigate aspects of pedagogy, ethics of place, and course and program assessment, this collection also offers voices "from the field," highlighting the experiences of faculty members, students, teaching assistants, and community members engaged in every aspect of off-campus study programs. Several chapters examine the programs in the traditional territories of Indigenous communities and in the Global South. Containing an appendix highlighting some examples of off-campus study programs, *Out There Learning* offers new pathways for faculty, staff, and college and university administrators interested in enriching non-traditional avenues of study.

DEBORAH CURRAN is an associate professor in the Faculty of Law and the School of Environmental Studies at the University of Victoria.

CAMERON OWENS is an associate teaching professor in the Department of Geography at the University of Victoria.

HELGA THORSON is an associate professor in and chair of the Department of Germanic and Slavic Studies at the University of Victoria.

ELIZABETH VIBERT is an associate professor in the Department of History at the University of Victoria.

OUT THERE LEARNING

CRITICAL REFLECTIONS ON OFF-CAMPUS STUDY PROGRAMS

Edited by

Deborah Curran

Cameron Owens

Helga Thorson

Elizabeth Vibert

With contributing editors

Matthew "Gus" Gusul

Duncan Johannessen

Kirsten Sadeghi-Yekta

UNIVERSITY OF TORONTO PRESS
Toronto Buffalo London

© University of Toronto Press 2019
Toronto Buffalo London
utorontopress.com
Printed in Canada

ISBN 978-1-4875-0411-3 (cloth)   ISBN 978-1-4875-2314-5 (paper)

∞

Printed on acid-free, 100% post-consumer recycled paper with
vegetable-based inks.

Library and Archives Canada Cataloguing in Publication

Out there learning : critical reflections on off-campus study
programs / edited by Deborah Curran, Cameron Owens, Helga
Thorson, Elizabeth Vibert ; with contributing editors Matthew
"Gus" Gusul, Duncan Johannessen, Kirsten Sadeghi-Yekta.

Includes bibliographical references and index.
ISBN 978-1-4875-0411-3 (cloth). ISBN 978-1-4875-2314-5 (paper)

1. University extension.   2. Service learning.   3. Community
and college.   I. Vibert, Elizabeth, 1962–, editor   II. Curran, Deborah,
1968–, editor   III. Owens, Cameron, 1970–, editor   IV. Thorson, Helga,
1964–, editor

LC6219.O98 2019   370.11'5   C2018-905845-5

This book has been published with the help of a grant from the Federation
for the Humanities and Social Sciences, through the Awards to Scholarly
Publications Program, using funds provided by the Social Sciences and
Humanities Research Council of Canada.

University of Toronto Press acknowledges the financial assistance to its
publishing program of the Canada Council for the Arts and the Ontario Arts
Council, an agency of the Government of Ontario.

Canada Council   Conseil des Arts
for the Arts   du Canada

ONTARIO ARTS COUNCIL
CONSEIL DES ARTS DE L'ONTARIO

an Ontario government agency
un organisme du gouvernement de l'Ontario

Funded by the   Financé par le
Government   gouvernement
of Canada   du Canada

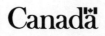

Canadä

MIX
Paper from
responsible sources
FSC® C016245

To our students and host communities,
for getting us out there.

Contents

List of Figures and Tables   x

Acknowledgments   xi

Introduction   3
HELGA THORSON AND MEGAN HARVEY

Section One: The Processes of Learning Out There

Where the Past and Present Intersect … Sam Kerr   25

1 "You Cannot Avoid All of This Past, Present, and Future
  When It's Everywhere Around You": Reflecting Relational
  Thinking in Field Study Experiences   27
  KACY MCKINNEY

Living in the Moment … Emily Halvorsen   47

2 An Integrative, Thematic Approach to International
  Field Study Programs   48
  AARON WILLIAMS

Being Part of Something Bigger … Kathleen O'Reilly   65

3 The Enlivened Classroom: Bringing the Field Back
  to Campus   66
  NAKANYIKE B. MUSISI

There Is No Front of the Classroom Here … Rob Cook   85

4 Settlers Unsettled: Using Field Schools and Digital
   Stories to Transform Geographies of Ignorance about
   Indigenous Peoples in Canada   87
   HEATHER CASTLEDEN, KILEY DALEY,
   VANESSA SLOAN MORGAN, AND PAUL SYLVESTRE

   Discovering Traces of the Past ... Sara Lax   107

Section Two: Implications of Place

   Connecting with the Community ... Aisling Kennedy   111

5 Outsider Education: Indigenous Law and
   Land-Based Learning   113
   JOHN BORROWS

   Live Life with Significance ... Freya Selander   134

6 Putting Law in Its Place: Field School Explorations of
   Indigenous and Colonial Legal Geographies   135
   DEBORAH CURRAN

   Mysteries Remain ... Laura Buchan   152

7 Power in Place: Dilemmas in Leading Field Schools
   to the Global South   153
   ELIZABETH VIBERT AND KIRSTEN SADEGHI-YEKTA

   What You Can't Get from a Textbook ... Sarah Elwood   174

Section Three: Assessing the Value of the Journey

   Eyes Wide Open ... Andrea van Noord   179

8 Getting Beyond "It Changed My Life": Assessment
   of Out There Transformation   181
   JANELLE S. PEIFER AND ELAINE MEYER-LEE

   9:14 AM: Saturday, 14 May 2016   202
   "The World Moves Through Us" ... Jake Noah Sherman   203

Contents   ix

9 Assessing Learning "Out There": Four Key Challenges
   and Opportunities 205
   CAMERON OWENS AND MARAL SOTOUDEHNIA

   Embracing Complexities ... Liah Formby 229

10 Transformation in the Field: Short-Term Study Abroad
    and the Pursuit of Changes 231
    MICHAEL R. GLASS

    Education Can Be Empowering ... Emily Tennent 252

    Concluding Remarks 253

Appendix: Field School Briefings 265
Contributors and Editors 279

Figures and Tables

Figures

2.1 The Cloaca Maxima, Tiber River, Rome   57
2.2 "Green capitalism is a lie"   61
8.1 Sample worksheet for assessing impacts in host community   188

Tables

8.1 Journeys Rubric with Selected Learning Objectives   198
9.1 A Summary of UBC's Sustainability Education Framework   219
10.1 Stages and Phases of Transformation   238

Acknowledgments

We have many people to thank for their contributions to this book. To begin with, we would like to thank Teresa Dawson, the former director of the Learning and Teaching Centre at the University of Victoria, for providing us with the means and the tools to embark on this project. She believed in the merits of this collaborative endeavour from the very start – and made it possible for us to achieve our goals. Through generous financial support from UVic's Learning and Teaching Centre in the form of three separate Learning Without Borders grants, the book moved from a random idea to a collective undertaking – one that not only resulted in a book publication, but also in forging a sense of community among colleagues who teach short-term off-campus study programs in our own city and beyond.

The initial grant allowed us to develop a field school forum on campus, bringing together faculty and staff from across the school who already run off-campus study programs, are interested in developing new programs, or provide support services (e.g., accounting, counselling, global engagement, risk management, etc.) for faculty running these programs. Building this sense of community from the ground up has made a tremendous difference on our campus. We would especially like to thank Brendan Burke and Silke Klenk for their initial contributions to the project. Through the second grant we were able to hire research assistants to help us dive into the research and to provide editorial assistance along the way. In particular, our gratitude goes out to Megan Harvey, Mariana Gallegos Dupuis, and Rebeca Macias Gimenez for their much-needed and much-appreciated research and editorial assistance. The final grant provided the means for us to invite additional authors from across Canada and the United States to campus to attend a workshop at the University of Victoria in March 2016. For this, we were able to hire Matthew "Gus" Gusul to organize the workshop and, with Kirsten Sadeghi-Yekta, to lead an applied theatre session with the book's authors, student participants in off-campus programs, and program teaching assistants. Many of the vignettes that appear in the book originated from that session. During the two-day workshop, the authors engaged in a field trip of our own. Reflecting on the importance of "place" in learning, we went on a hike around PKOLS Mountain with Kevin Paul, who told us stories about the traditional WSÁNEĆ (Saanich) territory before and after the arrival of the newcomers. We would like to express our appreciation to Kevin Paul for allowing us to learn through his stories.

Finally, we are honoured to have had the opportunity to work with an amazing group of authors who accepted our invitation to join the project. A special thank you to Michael Glass, Kacy McKinney, Elaine Meyer-Lee, Nakanyike Musisi, Janelle S. Peifer, and Aaron Williams for your fascinating contributions to the book, and to John Borrows, Heather Castleden, Kiley Daley, Vanessa Sloan Morgan, and Paul Sylvestre (and your original publishers) for agreeing to let us reprint what we consider to be groundbreaking articles in this book. We would like to express our gratitude to Gillian Calder, associate dean in the Faculty of Law at the University of Victoria, for her helpful and astute comments pertaining to one of the chapters. Above all else, we would like to thank the student contributors for their vignettes and, as our dedication emphasizes, all the students and host communities who have engaged with us in our off-campus study programs over the years. Without you, this book would not have been worth the effort.

OUT THERE LEARNING

Critical Reflections on Off-Campus Study Programs

Introduction

HELGA THORSON AND MEGAN HARVEY

When heading off to college or university for the first time, students may not realize that the most impactful learning experiences they may encounter might actually not occur on campus at all. Off-campus study programs such as global exchange, co-ops, internships, study abroad, travel tours, and field schools have the potential of opening students' eyes to the world around them, challenging their perspectives and world views as they gain knowledge, skills, and lifelong friendships. Within a broad framework of student mobility at the post-secondary level in general, this book focuses on one of these types of experiences: short-term off-campus programs for a small group of students led by at least one instructor from the home institution that focus on a specific topic of study in a location removed from the students' everyday lives. Traditionally called field schools[1] in disciplines such as archaeology, geography, or geology, short-term study abroad or study tours in foreign language disciplines, or global semesters on some college and university campuses, these off-campus programs are credit-based courses with specific learning goals.

This book developed out of an interest in investigating the pedagogical implications of these short-term off-campus programs among colleagues from various disciplines at the University of Victoria in British Columbia, Canada. After running field schools and short-term study tours ourselves, we were amazed by the educational quality and value we had experienced on our own programs and wanted to further investigate the academic impact as well as potential concerns of this type of learning. Inviting colleagues from across Canada and the United States, we hosted a workshop that explored the transformative potential of faculty-led off-campus study programs. This two-day workshop included a field trip to a nearby mountain, led by Kevin Paul, a poet and member of the WSÁNEĆ (Saanich) Nation, in which workshop participants were able to explore the land first-hand and hear stories of its history and people. In this way, we attempted to replicate, to a small degree, a field-learning experience into our workshop discussions. At the same time, the stories we heard, the conversations we had, and the land we walked on reminded us of the value of "place" in learning.

We all agree that this type of learning has the potential to be life changing for students, teaching assistants, and instructors alike, and we wanted to know how and why. What was it about these types of programs that facilitated deep learning? Could these learning experiences be transferred back into a more traditional classroom setting, influencing all of our teaching in new and exciting ways? Or is the learning that happens on these programs context specific, allowing a hands-on experiential component that is hard to replicate in a classroom? Are there aspects of field schools or study tours that actually lead to bad pedagogical practices? Are they exclusionary, not only in the financial sense, but also in terms of ability or mindset? Could they actually hinder cross-cultural learning, reflection, and dialogue by putting participants in situations in which they remain enmeshed in their own cultural context rather than becoming immersed in the communities they set out to study? Do these programs have a long-term impact on students' lives? These questions and others form the basis of this book.

Out There Learning: Critical Reflections on Off-Campus Study Programs explores the pedagogical foundations of what it means to learn "out there" – in the field, so to speak – and simultaneously investigates why this type of learning, especially when compared with traditional classroom settings, can sometimes be viewed as "out there": as different, erratic, or even foolish. This book is neither envisioned as a how-to guide nor as a celebration of exciting trips and amazing group experiences; it is an informed critical interrogation of pedagogical philosophies and practices that brings together the authors' individual research as well as their own experiences teaching or administering short-term off-campus study programs. The types of programs the authors run vary from one another in terms of course topic; geographical regions visited; amount of travelling entailed; size of the group; numbers of faculty involved; whether students share accommodations with one another or stay with home-stay families; whether the course takes place exclusively off-campus or whether the time spent off-campus is integrated within a longer on-campus course; and length of the program, among other things. What we mean by "short-term" ranges from a one-week global journey offered over spring break (see the chapters by Musisi and by Peifer and Meyer-Lee in this volume) to a three-month program (see McKinney's chapter).[2] Although the type and length of the programs discussed vary widely, the contributions to this book seek to discover what, if anything, makes these programs different from on-campus classroom learning. Based on critical reflection, *Out There Learning* challenges assumptions about pedagogy, about place, and about personal transformation in off-campus educational settings, and makes recommendations for getting the most out of faculty-led off-campus study programs.

The first section of the book, "The Processes of Learning Out There," investigates the pedagogy of short-term off-campus programs and their transformative potential for students, instructors, and the communities involved. It focuses on learning that is "out there" in the double sense of the term, on processes of unlearning, and on the journey that is undertaken that is typically also a journey of the self. How can these off-campus study programs be set up to increase the transformative learning potential for all participants? In what ways is learning on a field school or study tour different from what goes on in a traditional classroom?

The second section, "Implications of Place," examines what happens when we leave campus and study in another location and in communities significantly different from our own. How does this sense of place affect our experiences, emotions, and our conception of who we are as well as our relationship to the world around us? Is a "different" place necessary for transformative learning? How can instructors ensure that student experience in a different place does not reinforce stereotypes? Does community engagement make a difference in the program participants' lives and the lives of those who are a part of the communities we visit? This section of the book focuses on the importance of "place" and "space" in the learning and teaching of off-campus study programs, and it highlights the reciprocal relationship and engagement with the communities involved; it does not, however, measure or evaluate the impacts on these communities. This important topic is beyond the scope of this volume.

The third section of the book, "Assessing the Value of the Journey," examines the outcomes and assessment of field schools and short-term off-campus study programs. How do we know that these programs are as transformative as we imagine them to be? In what ways can the impact of field schools be measured? To what extent does this type of learning continue after the program is over? What are the learning outcomes and after-effects for students, faculty, and teaching assistants? How do instructors' understandings of the outcomes differ from those of students, parents, administrators, and others? Are there ways to help students articulate the learning outcomes and career competencies that stemmed directly from their off-campus study experiences? What are the beneficial or detrimental outcomes of field programs for host communities and/or the students' home communities? This section not only offers useful assessment strategies, resources, and examples (as in the chapters by Peifer and Meyer-Lee and by Owens and Sotou- dehnia), but also critical reflections on what precisely we can and can- not measure (see Owens and Sotoudehnia as well as Glass) and whether transformation should even serve as a goal of short-term off-campus programs (Glass).

Besides individual chapters investigating aspects of pedagogy, the ethics of place, and assessment, we have included "intermezzos" throughout the book that highlight voices "from the field." These vignettes interspersed between sections and chapters provide insights based on lived experience, uncovering memorable moments from var- ious off-campus study programs. They are deliberately not matched to their relevant field schools, our thinking being that these short nar- ratives speak across themes, theoretical discussions, and experiences. Vignettes not only foreground voices of students, faculty, community members, and teaching assistants, but also illustrate the "messiness" of learning by uncovering learning processes that can sometimes be uncomfortable and unsettling – but also deeply enriching. Finally, we end the book with reflections from all of the authors and editors, in dialogue with one another, about their experiences teaching and administering off-campus study programs and distinct areas in need of future research.

Four threads run throughout the book. These include (1) field schools and short-term off-campus study programs as curated learning experiences, (2) the importance of disciplinary frameworks and perspec- tives that have led to discipline-specific traditions and histories regard- ing off-campus student learning and opportunities for transcending these boundaries, (3) the transformative potential of these off-campus learning opportunities, and (4) the intricate relationships that develop through community engagement.

Interwoven Threads and Common Themes

Thread 1: Curated Learning Experience

Just as an art gallery or museum curates an exhibit, instructors and directors of off-campus study programs sift through material and decide what to present as they create their course outlines and travel itineraries. Similar to visitors of a museum or art gallery, students come to the course with various degrees of background knowledge, different ways of approaching the material, and their own sets of expectations, opinions, perspectives, and emotions. Moreover, short-term off-campus study programs are full-time endeavours. Students do not have other immediate distractions such as work or family obligations like they do back home. Obviously not everything or every minute can be planned and organized; spontaneity and chance encounters are also very much a part of field schools and study tours and should be both an expected and a valued part of the learning process. At its very core, this book explores the implications of short-term off-campus study programs as curated experiences and how these experiences affect the learning process. Instructors leading off-campus programs, for example, make choices in terms of the overall itinerary, the places that are visited (or avoided) in each location, the degree to which they would like to see participants immerse themselves in the local community, and the narrative they implicitly or explicitly want the students to construct – and these choices have implications that affect the overall student experience (see McMorran 2015). Each of the chapters in this book addresses the curated nature of field schools, whether in terms of pedagogy, relationship to place, or program assessment.

Thread 2: Cross-Disciplinary and Interdisciplinary Perspectives

Most of the research on short-term off-campus study programs is discipline specific.[3] Some scientific disciplines have traditionally viewed fieldwork as an extension of the laboratory – where the laboratory is an actual "field" where scientists excavate rock samples or traces of ancient civilizations. Disciplines such as anthropology and geography have a long tradition of field schools, whereas others, such as law or history, do not. Law schools, for example, typically emphasize clinical experiences outside of the classroom rather than field schools or study tours, and students of history traditionally spend their time researching in a library or archive.

Although there are exceptions, off-campus study programs in these disciplines are quite recent additions to the undergraduate or graduate curriculum. Language departments, in contrast, have a long tradition of study abroad programs, and much research has been conducted on short-term study abroad programs that have a language-learning focus.[4]

This book examines the topic of short-term off-campus programs across disciplines and recognizes that each discipline has its own unique history and set of discipline-specific learning goals. Yet, we also realize that there are aspects of field schools and off-campus study tours that are similar across disciplines. While recognizing the differences, it is precisely these points of convergence that we want to emphasize in this book. We acknowledge that short-term off-campus study programs can be quite interdisciplinary in nature. Some programs are designed so that the course is co-taught by instructors from different disciplines, other programs are based on interdisciplinary course content, and still others admit students from multiple disciplines. Attention to "place" that comes from being "somewhere else" often relies on an interdisciplinary approach. Short-term off-campus study programs vary widely from one another.[5] The objective of this book is to examine them in a cross-disciplinary as well as an interdisciplinary way.[6]

Thread 3: The Transformative Potential of Short-Term Off-Campus Programs

According to Patricia Cranton, "Transformative learning has to do with making meaning out of experiences and questioning assumptions based on prior experience" (2006, 8). Jack Mezirow introduced the concept of transformative learning in adult learners in 1975 and continued to develop his theories over the next several decades. Transformative learning processes involve seeing things from other perspectives and changing "habits of mind" (Mezirow 1997, 5). Getting out of one's comfort zone through a change in routine and location combined with rigorous academic inquiry and reflection helps facilitate a process by which learners can question and re-evaluate their assumptions and values. A strong focus of this book is to examine how best to facilitate this process so that students, instructors, and teaching assistants get the most out of their off-campus experiences.

Although field schools and study tours are often based on experiential learning, they are not necessarily one and the same. It is possible, for example, to study in a different location and engage in classroom behaviour that is not experiential in nature. The same can be said for the often-touted transformative nature of short-term off-campus study programs. Participating in an off-campus study program can be life changing for students – but the transformative nature of field schools and study tours is not a guarantee. This book reflects on what "transformative learning" means in the context of off-campus study programs and investigates learning and teaching techniques that could possibly enhance the educational experience of these courses, thereby increasing their potential for transformative learning. Some of these techniques include the following:

Affect and emotion: often referred to as holistic learning or empathy-based instruction, this approach focuses on "the whole person." It is not concerned exclusively with expanding the intellect but rather it includes an integration of the intellectual, social, and emotional sides of the learner (see Love and Love 1995; Boler 1999; Berman 2004; Mitussis and Sheehan 2013).

Dialogue and storytelling: reflection occurs not only in isolation, but also in dialogue with others (see Freire 1970; Baker, Jensen, and Kolb 2002; Alexander 2006); related to this is the power of narrative and storytelling (see Alterio and McDrury 2004; Archibald 2008; Gottschall 2012).

Experiential learning: learners are given "hands-on" tasks or put in a setting in which they can apply their skills, knowledge, and feelings (see Dewey 1997 [1938]; Kolb 1984).

Reflection: experiential learning is not enough in and of itself; according to John Dewey, we do not learn from experience per se but rather from reflecting on experiences (Kolis 2013, 103; see also Mezirow and Associates 1990).

Social action: transformative learning has the potential to lead one towards social change, to jolt a learner into action (see hooks 1994; Fielding 2001; Brookfield 2005).

Out There Learning: Critical Reflections on Off-Campus Study Programs is interested in exploring and analysing the techniques, environments, and dimensions that could enhance the transformative learning potential of field schools and off-campus study tours for students, instructors, teaching assistants, and to a lesser extent for community members who have interacted and worked with course participants.

Thread 4: Community Engagement

A final strand woven throughout this book is community engagement. On one level, we discuss techniques that can be used to facilitate an open, engaged, and reciprocal relationship between community members and course participants. On another level, the authors of this book investigate the intricacies of community engagement and relationship building, including ways in which we might be doing more harm than good.

In her chapter in the book *The World Is My Classroom: International Learning and Canadian Higher Education*, Joanne Benham Rennick discusses what it means to be a good global citizen in the Canadian educational context, particularly in the realm of service learning. She maintains that "unless we establish and embed particular and explicit values in our programming, we are likely to perpetuate a neo-colonial agenda that carries a subtext of 'saving,' 'helping,' and even 'civilizing' partners in what is now sometimes called the global South" (2013, 37). Rebecca Tiessen and Robert Huish, the editors of *Globetrotting or Global Citizenship? Perils and Potential of International Experiential Learning* (2013), similarly point out the ethical dilemmas involved in studying, volunteering, or completing a work-experience term abroad, particularly in the Global South. They point out that these experiential learning opportunities may be more about "globetrotting" than about cultivating global citizenship and challenging inequalities in the context of international development. Tiessen and Huish suggest that these programs "provide rich opportunities for understanding the causes of inequality and finding ways to work in solidarity with our partners in the Global South to challenge and circumvent structures of inequality" (4), and they highlight the need for a critical reflection on international experiential learning.

Out There Learning takes a critical perspective on community engagement initiatives on faculty-led short-term off-campus study programs. What happens when programs that traditionally have not formed relationships with communities such as those in earth and ocean sciences, geology, or archaeology begin to reach out to and learn from local communities? How can cross-cultural dialogues affect learning when students are put in situations where they are expected to discuss critical issues with their counterparts in other countries or members of the host communities? How can knowledge, awareness, and perspectives change when engaging with First Nations communities? What might communities gain from interacting with post-secondary students?

The process of writing the final chapter of this volume put authors in conversation with one another to discuss the writing process and to tease out some of the commonalities and tensions that emerged over the course of our interactions with one another. Two main questions arose in our informal conversation: To what extent do we view classroom learning and field schools as binary categories, separate from one another and incommensurable? To what extent is it possible to measure transformation in a short-term off-campus study program? It is clear that all of us were energized through our collaborations on this book project, and we hope that our critical reflections help energize and inspire you, the book's readers, as you contemplate what it means to learn and teach "out there" beyond the walls of the on-campus classroom.

What Does the Research Tell Us?

There has been much research and discussion pertaining to student mobility in higher education in general. Although it is outside of the scope of this book, we would like to acknowledge the traditions of practical learning and teaching in a variety of disciplines, including programs that have a long history of practicum placements, clinical training, co-op positions, and other hands-on professional training experiences in fields such as medicine, education, social work, business, and law that get the students out of the classroom and learning "out there."[7] Through a long history of scholarship contemplating the value of bridging the gap between theory and practice in these and other disciplines, as well as emerging studies in progressive pedagogy, interdisciplinary learning and teaching, transformative education, postcolonial studies, and critical feminist pedagogy, much research has emerged about ways to connect theoretical and practical skills, content knowledge and experiential learning, and classroom learning and community-engaged work and service. The chapters in this book address this body of literature in various ways.

Students face a wide variety of choices and opportunities that lead them from the classroom out into the world, including opportunities for vocational education; co-op, practicum, and internship placements; language and cultural learning through study abroad experiences; service learning; and international education, as well as their own personal travel. There is much research on these types of opportunities. The focus of this section of the introduction, however, as well as of this book, is on short-term off-campus programs. As a reminder, we define these as courses, parts of courses, or series of courses offered by colleges or universities for a small group of students led by at least one instructor from the home institution that focuses on a specific topic of study in a location removed from the students' everyday lives. These programs are often called "field schools" or "study tours," but the names can vary depending on the region, the discipline, or the specific institution offering these short-term off-campus study programs.

Until relatively recently, much of the published work on field schools and off-campus study tours could be characterized as self-promotional and descriptive. For the most part, those writing about short-term off-campus programs have done so from a personal and professional conviction in the overwhelming benefits of the model for students, instructors, institutions, and, where applicable, for community partners. Authors are generally excited about off-campus study programs and what they have achieved, and wish to promote their value while offering some knowledge sharing from their experiences in order to assist and encourage other current or would-be program organizers.

The typically short pieces that characterize this literature tend to cover a broad spectrum of information in each article or chapter, including descriptions of the structure of the program (duration, location, number of students, ratio of instructors to students), practical/logistical issues, problem-solving techniques, student feedback, and an argument for the value of off-campus study programs for specific disciplines or for postsecondary education generally (see Deiparine 1987; Gmelch and Gmelch 1999; Grant, Preissle, Beoku-Betts, Finlay, and Fine 1999; Iris 2004; Ward 1999). Although these pieces are engaging and useful reading material for those involved in running field schools, there is an identifiable need for research that offers theoretically informed and critical analyses of the pedagogies and practices of field schools. These have the potential to offer "information exchange" at a higher level of understanding of what it is, exactly, that off-campus study tours contribute to student learning. It is the desire for such exchanges that inspired this volume.

Within the current literature on field schools and short-term off-campus study tours, several debates have emerged that are particularly useful in understanding the pedagogy and practices of these programs: Are short-term off-campus study programs exclusive? What is the purpose of field schools and study tours? What makes field schools and study tours different from other in-class learning? Through their contributions in *Out There Learning: Critical Reflections on Off-Campus Study Programs*, the authors draw on and expand these debates from a cross-disciplinary and interdisciplinary perspective.

Are Short-Term Off-Campus Study
Programs Exclusive?

Where lively debate about field schools or off-campus study tour programs exists, it tends to emerge from disciplines with long traditions of field study and be contained within discipline-specific journals or book collections. For example, the issue of inclusivity and field schools appears to be almost entirely restricted to geography, although it could certainly apply more broadly.[8] Karen Nairn's (1996; 1999; 2005) is the most frequently cited critical voice in this debate. She identifies the underlying normative assumptions operating in the ways geography field schools are conceived, practised, and experienced, pointing to the gendered and ableist dimensions of the emphasis on a "healthy" or "fit" physical body required for field work (Nairn 1996; 1999).

Tim Hall, Mick Healey, and Margaret Harrison (2002) have also addressed this issue, recognizing that geography fieldwork features some specific and potentially prohibitive challenges for students with disabilities. They assert that access to fieldwork for people of various abilities must be considered when evaluating the effectiveness of fieldwork (or field schools) as a mode of learning (215). While directed at the "rugged terrain" typical of geography field schools (and perhaps archaeology field schools), these authors' observations could apply to any non-local, residential, off-campus study tour. If some exclusion on the basis of ability cannot be avoided, these authors challenge us to rethink the field school model in such a way that at least some disabilities be perceived as opportunities to offer alternative insights rather than acting as barriers to student participation in experiential learning (221).

Clare Herrick (2010) looks at inclusivity from the somewhat different angle of cost. She notes that fieldwork has become more difficult to fund in a period of recent public cutbacks in the United Kingdom, with fees being displaced onto students. The effect has been a shift in student expectations towards getting "value for money" from their fieldwork experiences (see also Kent, Gilbertson, and Hunt 1997). Interestingly, Thomas Bordelon and Iris Phillips's (2006) study of student perceptions of service learning found that when students were paid wages as part of their service learning experience, their overall satisfaction with the program decreased.

The potential benefits of off-campus study programs for student learning are frequently cited, and such programs are often promoted as models of transformative education. Less common are works that articulate the potential risks of off-campus study tours for student learning. In addition to the concerns around reinforcing a normative body as gendered male (Nairn 1996) and physically "fit" (Nairn 1999), Nairn (2005) warns that uncritical promotions of field schools as offering students "direct experience" of the world could confirm rather than unsettle their prejudicial preconceptions.[9] Overall, the debates over the inherent value of off-campus student fieldwork courses in post-secondary geography curricula appear to have inspired more thoughtful arguments *for* the continued value of field study rather than putting field study programs at risk. Moreover, these arguments have been made stronger by taking critical perspectives into account (Dunphy and Spellman 2009; Fuller et al. 2006; Herrick 2010; Hope 2009; Owens, Sotoudehnia, and Erickson-McGee 2015).

What Is the Purpose of Field Schools and Study Tours?

In anthropology, a discipline with a long critical tradition concerning the politics of power in ethnography, recent writing on field schools reflects an increasing emphasis on community collaboration, through which off-campus study tours strive to model research relationships in which communities are equal partners if not directors or co-directors (see Beck 2006; Carlson, Lutz, and Schaepe 2009; Hyatt et al. 2011; Menzies 2004; Menzies and Butler 2011).[10] These authors recognize that although terms such as "collaboration" and "community-based research" are on the rise – particularly in relation to research with Indigenous communities – creating and sustaining field schools that model collaborative research relationships can take time to develop, care to maintain, and is a worthy topic of debate in the literature.

Some voices in the discipline question whether this focusing of ethnographic field schools on equalizing power relationships between researchers and communities occurs at the expense of students receiving rigorous training in the research techniques that define the discipline (Wallace 2011; Hyatt et al. 2011; Menzies and Butler 2011). The framing of this as an either/or scenario may signal a backlash against the increasingly prominent language of relationship building (forming community partnerships), often associated with transformative learning. Based on the literature, an argument could be made that both approaches are about developing a set of transferable skills in students, but there is some tension between advocates for the "hard" skills of methodological training (in this case, ethnography) versus the "soft" skills of community engagement (e.g., communication skills, working in a team, problem solving, etc.); see Cobb and Croucher (2012). Tim Wallace (2011) finds that in attempts to balance student training with the needs of participating communities, student learning is often sacrificed, leaving students underskilled and undertrained. Beyond anthropology, however, other scholars see no issue to debate. Rather, they acknowledge that in an "increasingly market-driven academy" transferable skills and training are a practical necessity (Fuller et al. 2006),[11] and recognize that the soft and hard skills field schools provide are both highly desirable in the marketplace (Castleden et al. 2013, reprinted as chapter 4 of this volume; Cobb and Croucher 2012; Fuller et al. 2006; Kent, Gilbertson, and Hunt 1997, 319; Mitussis and Sheehan 2013, 45).

What Makes Field Schools and Study Tours Different from Other In-Class Learning?

Several authors mention (Castleden et al. 2013; Dunphy and Spellman 2009; Elkins and Elkins 2007; Fuller et al. 2006; Gmelch and Gmelch 1999; Herrick 2010, 177; Sheffield 2015), but few explore in depth (Lawrence 2008; Mitussis and Sheehan 2013), the role of affect in the experiential and transformative learning that short-term off-campus study programs provide. Certainly, affect seems to be an important dimension of the "deeper learning" promoted in such courses, but it is underexplored and undertheorized in the field school literature.

The study conducted by Fuller et al. (2006, 94) suggests that fieldwork is an effective learning experience partly because students find it enjoyable, which is largely attributable to the increased student contact with instructors characteristic of many field schools. Gmelch and Gmelch (1999) arrive at a different conclusion: they surmise it is the experience of student discomfort (in their case, a product of culture shock) that stimulates maturation and personal growth, which is what makes field school experiences so rewarding for students. More recently, Herrick (2010) argues for the pedagogical value of risk in fieldwork, which instructors and university administrations work so hard to reduce or eliminate. When students are confronted with "real" consequences to their actions, they must adapt, and it is partly this adaptation to changing circumstances that makes a learning experience transformative (114).

Mitussis and Sheehan (2013, 50) make a similar observation, although in their formulation, not all discomfort is productive. They make a distinction between "constructive" and "destructive" emotional friction but argue that when appropriately integrated into the learning experience, the "visceral nature" of emotional experiences can foster critical moments for transformative learning (50–1). The intensity of the learning experience offered through short-term off-campus study programs is partly what makes them so effective, as students are "immersed in their learning, rather than confronted by it at intervals in the lecture room or during assessment" (50–1). Notwithstanding these insights, much remains to be explored about the complex role of affect in student learning through off-campus study tours.

Bringing Forward Multidisciplinary Perspectives

As already mentioned, the multidisciplinary contributions in this volume bring together insights and debates that have, until now, largely been isolated within individual disciplines. Moreover, this collection seeks to contribute increased disciplinary diversity to the published work on field schools and off-campus study terms. Where research publications on off-campus study programs have widespread representation in fields such as geography, environmental sciences, archaeology, and sociocultural anthropology as well as foreign language disciplines that study linguistic and cultural immersion, the range of disciplines now offering off-campus study programs as part of their curricula is much more extensive. Increased representation of this diversity in scholarly publications has the potential to bring to the fore fresh perspectives on the possibilities and challenges of field schools and other off-campus study programs, which will get us "out there" learning in the field in new and interesting ways.

NOTES

1 The term "field school" is used much more commonly across disciplines in Canada, while in the United States it is usually reserved for field-based programs in the sciences and social sciences. In some contexts, "field school" is used to describe an off-campus course taught in one location and a "study tour" is an off-campus course that involves travelling to two

or more locations; however, this is not always the case. For the purpose of this book, we use the term "field school" interchangeably with the term "off-campus study program" because that is how the terms are understood at the editorial team's home institution.

2 Whereas Glass (chapter 10) defines "short term" as less than four weeks, the editors of this volume realize that some institutions define this differently. Therefore, we have included discussions of programs up to three-months in length in this book, although most are indeed less than four weeks in length.

3 This does not imply that discipline-specific research is not valuable to other fields of study. Moreover, research on specific topics that emerge out of one or two specific disciplines, such as France et al.'s (2015) book on the use of mobile technology during field schools, can be quite useful in a range of educational contexts and scenarios. *Out There Learning* does not delve into the use of technology and social media during short-term off-campus programs. See, France et al. (2015) for more on mobile learning.

4 See, for example, journals dedicated specifically to international education and study abroad such as *Frontiers: International Journal of Study Abroad* or the *Journal of Studies in International Education*.

5 Some field schools, for example, focus on a particular site and others visit multiple sites with resultant trade-offs negotiated between depth and breadth and whether the focus is on a single-case study or a comparative analysis.

6 We would like to acknowledge two books that have served as models for us when we began to write our book: *The World Is My Classroom: International Learning and Canadian Higher Education* (2013), edited by Rennick and Desjardins, and *Learning and Teaching Community-Based Research: Linking Pedagogy to Practice* (2014), edited by Etmanski, Hall, and Dawson. *Putting the Local in Global Education: Models for Transformative Learning Through Domestic Off-Campus Programs* (2015), edited by Neal W. Sobania, is another excellent cross-disciplinary collaboration.

7 Journals such as the *Journal of Teacher Education,* founded in 1950, the *Journal of Cooperative Education and Internships,* founded in 1965, *Medical Education,* founded in 1966, and *Action in Teacher Education,* founded in 1978, point to the long-standing traditions of practical education in certain disciplines.

8 A recent exception is Clarke and Philips (2012).

9 For a response, see Hope (2009).

10 For a prescriptive vision of collaborative archeology, see La Salle (2010).

11 Although not about field schools per se, Harry Arthurs (2013, 10) cautions his colleagues in law that the forms of experiential learning offered through law schools "can degenerate into mere skills training" rather than offer students opportunities to witness the complex practices of law.

REFERENCES

Alexander, Robin. 2006. *Towards Dialogic Teaching: Rethinking Classroom Talk.* 3rd ed. York: Dialogos.

Alterio, Maxine, and Janice McDrury. 2004. *Learning through Storytelling in Higher Education: Using Reflection and Experience to Improve Learning.* London: Taylor and Francis e-Library.

Archibald, Jo-ann [Q'UM Q'UM XIIEM]. 2008. *Indigenous Storywork: Educating the Heart, Mind, Body, and Spirit.* Vancouver: UBC Press.

Arthurs, Harry W. 2013. "The Future of Legal Education: Three Visions and a Prediction." *Osgoode Law School Comparative Research in Law & Political Economy Research Series.* Research Paper No. 49: 1–13. Accessed 24 Feb. 2017. http://digitalcommons.osgoode.yorku.ca/clpe/291. https://doi.org/10.2139/ssrn.2349633.

Baker, Ann C., Patricia J. Jensen, and David A. Kolb. 2002. *Conversational Learning: An Experiential Approach to Knowledge Creation.* Westport, CT: Quorum Books.

Beck, Sam. 2006. "Community Service-Learning: A Model for Teaching and Activism." *North American Dialogue* 9 (1): 1–7. https://doi.org/10.1525/nad.2006.9.1.1.

Berman, Jeffrey. 2004. *Empathic Teaching: Education for Life.* Amherst: University of Massachusetts Press.

Boler, Megan. 1999. *Feeling Power: Emotions and Education.* New York: Routledge.

Bordelon, Thomas D., and Iris Phillips. 2006. "Service-Learning: What Students Have to Say." *Active Learning in Higher Education* 7 (2): 143–53. https://doi.org/10.1177/1469787406064750.

Brookfield, Stephen. 2005. *The Power of Critical Theory: Liberating Adult Learning and Teaching.* San Francisco: Jossey-Bass.

Carlson, Keith, John Lutz, and David Schaepe. 2009. "Turning the Page: Ethnohistory from a New Generation." *University of the Fraser Valley Research Review* 2 (2): 1–8.

Castleden, Heather, Kiley Daley, Vanessa Sloan Morgan, and Paul Sylvestre. 2013. "Settlers Unsettled: Using Field Schools and Digital Stories to Transform Geographies of Ignorance about Indigenous Peoples in Canada." *Journal of Geography in Higher Education* 37 (4): 487–99. https://doi.org/10.1080/03098265.2013.796352.

Clarke, Amanda, and Tim Phillips. 2012. "Archaeology for All? Inclusive Policies for Field Schools." In *Global Perspectives on Archaeological Field Schools: Constructions of Knowledge and Experience,* ed. Harold Mytum, 41–59. New York: Springer. https://doi.org/10.1007/978-1-4614-0433-0_4.

Cobb, Hannah, and Karina Croucher. 2012. "Field Schools, Transferable Skills and Enhancing Employability." In *Global Perspectives on Archaeological Field Schools: Constructions of Knowledge and Experience*, ed. Harold Mytum, 25–40. New York: Springer. https://doi.org/10.1007/978-1-4614-0433-0_3.

Cranton, Patricia. 2006. *Understanding and Promoting Transformative Learning: A Guide for Educators of Adults.* 2nd ed. San Francisco: Jossey-Bass.

Deiparine, Marilou. 1987. "The First Inter-institutional Field School in Prehistoric Archaeology." *Philippine Quarterly of Culture and Society* 15 (3): 255–6.

Dewey, John. 1997 [1938]. *Experience and Education.* New York: Touchstone.

Dunphy, Alison, and Greg Spellman. 2009. "Geography Fieldwork, Fieldwork Value and Learning Styles." *International Research in Geographical and Environmental Education* 18 (1): 19–28. https://doi.org/10.1080/10382040802591522.

Elkins, Joe, and Nichole M.L. Elkins. 2007. "Teaching Geology in the Field: Significant Geoscience Concept Gains in Entirely Field-Based Introductory Geology Courses." *Journal of Geoscience Education* 55 (2): 126–32. https://doi.org/10.5408/1089-9995-55.2.126.

Etmanski, Catherine, Budd L. Hall, and Teresa Dawson, eds. 2014. *Learning and Teaching Community-Based Research: Linking Pedagogy to Practice.* Toronto: University of Toronto Press.

Fielding, Michael. 2001. "Students as Radical Agents of Change." *Journal of Educational Change* 2 (2): 123–41. https://doi.org/10.1023/A:1017949213447.

France, Derek, W. Brian Whalley, Alice Mauchline, Victoria Powell, Katherine Welsh, Alex Lerczak, Julian Park, and Robert Bednarz. 2015. *Enhancing Fieldwork Learning Using Mobile Technologies.* New York: Springer. https://doi.org/10.1007/978-3-319-20967-8.

Freire, Paulo. 1970. *Pedagogy of the Oppressed.* New York: Continuum Books.

Fuller, Ian, Sally Edmondson, Derek France, David Higgitt, and Ilkka Ratinen. 2006. "International Perspectives on the Effectiveness of Geography Fieldwork for Learning." *Journal of Geography in Higher Education* 30 (1): 89–101. https://doi.org/10.1080/03098260500499667.

Gmelch, George, and Sharon Bohn Gmelch. 1999. "An Ethnographic Field School: What Students Do and Learn." *Anthropology & Education Quarterly* 30 (2): 220–7. https://doi.org/10.1525/aeq.1999.30.2.220.

Gottschall, Jonathan. 2012. *The Storytelling Animal: How Stories Make Us Human.* New York: Houghton Mifflin Harcourt.

Grant, Linda, Judith Preissle, Josephine Beoku-Betts, William Finlay, and Gary Alan Fine. 1999. "Fieldwork in Familiar Places: The UGA Workshop in Fieldwork Methods." *Anthropology & Education Quarterly* 30 (2): 238–48. https://doi.org/10.1525/aeq.1999.30.2.238.

Gronski, Robert, and Kenneth Pigg. 2000. "University and Community Collaboration: Experiential Learning in Human Services." *American Behavioral Scientist* 43 (5): 781–92. https://doi.org/10.1177/00027640021955595.

Hall, Tim, Mick Healey, and Margaret Harrison. 2002. "Fieldwork and Disabled Students: Discourses of Exclusion and Inclusion." *Transactions of the Institute of British Geographers* 27 (2): 213–31. https://doi.org/10.1111/1475-5661.00050.

Herrick, Clare. 2010. "Lost in the Field: Ensuring Student Learning in the 'Threatened' Geography Fieldtrip." *Area* 42 (1): 108–16. https://doi.org/ 10.1111/j.1475-4762.2009.00892.x.

hooks, bell. 1994. *Teaching to Transgress: Education as the Practice of Freedom.* New York: Routledge.

Hope, Max. 2009. "The Importance of Direct Experience: A Philosophical Defense of Fieldwork in Human Geography." *Journal of Geography in Higher Education* 33 (2): 169–82. https://doi.org/10.1080/03098260802276698.

Hyatt, Susan, Marcela Castro Madariaga, Margaret Baurley, Molly J. Dagon, Ryan Logan, Anne Waxingmoon, and David Plasterer. 2011. "Walking the Walk in Collaborative Fieldwork: Responses to Menzies, Butler, and Their Students." *Collaborative Anthropologies* 4 (1): 243–51. https://doi.org/10.1353/ cla.2011.0017.

Iris, Madelyn. 2004. "Fulfilling Community Needs through Research and Service: The Northwestern University Ethnographic Field School Experience." *NAPA Bulletin* 22 (1): 55–71. https://doi.org/10.1525/napa.2004.22.1.055.

Kent, Martin, David D. Gilbertson, and Chris O. Hunt. 1997. "Fieldwork in Geography Teaching: A Critical Review of the Literature and Approaches." *Journal of Geography in Higher Education* 21 (3): 313–32. https://doi.org/ 10.1080/03098269708725439.

Kolb, David A. 1984. *Experiential Learning: Experiences as the Source of Learning and Development.* Upper Saddle River, NJ: Prentice Hall.

Kolis, Mickey. 2013. *Rethinking Teaching: Classroom Teachers as Collaborative Leaders in Making Learning Relevant.* Lanham, MD: Rowman and Littlefield.

La Salle, Marina J. 2010. "Community Collaboration and Other Good Intentions." *Archaeologies* 6 (3): 401–22. https://doi.org/10.1007/s11759-010-9150-8.

Lawrence, Randee Lipson. 2008. "Powerful Feelings: Exploring the Affective Domain of Informal and Arts-Based Learning." *New Directions for Adult and Continuing Education* 120 (1): 65–77. https://doi.org/10.1002/ace.317.

Love, Patrick G., and Anne Goodsell Love. 1995. "Enhancing Student Learning: Intellectual, Social, and Emotional Integration." *ASHE-ERIC Higher Education Report* 24 (4): 1–5.

McMorran, Chris. 2015. "Between Fan Pilgrimage and Dark Tourism: Competing Agendas in Overseas Field Learning." *Journal of Geography*

in Higher Education 39 (4): 568–83. https://doi.org/10.1080/03098265.2015.
1084495.

Menzies, Charles R. 2004. "Putting Words into Action: Negotiating Collaborative
Research in Gitxaala." *Canadian Journal of Native Education* 28 (1/2): 15–32.

Menzies, Charles R., and Caroline F. Butler. 2011. "Collaborative Service
Learning and Anthropology with Gitxaała Nation." *Collaborative Anthropologies*
4 (1): 169–242. https://doi.org/10.1353/cla.2011.0014.

Mezirow, Jack. 1997. "Transformative Learning: Theory to Practice." *New Directions
for Adult and Continuing Education* 74: 5–12. https://doi.org/10.1002/ace.7401.

Mezirow, Jack, and Associates. 1990. *Fostering Critical Reflection in Adulthood: A
Guide to Transformative and Emancipatory Learning.* San Francisco: Jossey-Bass.

Mitussis, Darryn, and Jackie Sheehan. 2013. "Reflections on the Pedagogy
of International Field-schools: Experiential Learning and Emotional
Engagement." *Enhancing Learning in the Social Sciences* 5 (3): 41–54. https://
doi.org/10.11120/elss.2013.00013.

Nairn, Karen. 1996. "Parties on Geography Fieldtrips: Embodied Fieldwork?"
NZ Women's Studies Journal 12 (2): 86–97.

Nairn, Karen. 1999. "Embodied Fieldwork." *Journal of Geography* 98 (6): 272–82.
https://doi.org/10.1080/00221349908978941.

Nairn, Karen. 2005. "The Problems of Utilizing 'Direct Experience' in Geography
Education." *Journal of Geography in Higher Education* 29 (2): 293–309. https://doi.
org/10.1080/03098260500130635.

Owens, Cameron, Maral Sotoudehnia, and Paige Erickson-McGee. 2015.
"Reflections on Teaching and Learning for Sustainability from the Cascadia
Sustainability Field School." *Journal of Geography in Higher Education* 39 (3):
313–27. https://doi.org/10.1080/03098265.2015.1038701.

Rennick, Joanne Benham. 2013. "Canadian Values, Good Global Citizenship,
and Service Learning in Canada: A Socio-Historical Analysis." In *The World Is
My Classroom: International Learning and Canadian Higher Education*, ed. Joanne
Benham Rennick and Michel Desjardins, 23–44. Toronto: University of
Toronto Press. https://doi.org/10.3138/9781442669079-004.

Rennick, Joanne Benham, and Michel Desjardins, eds. 2013. *The World Is My
Classroom: International Learning and Canadian Higher Education.* Toronto:
University of Toronto Press. https://doi.org/10.3138/9781442669079.

Sheffield, Eric C. 2015. "Toward Radicalizing Community Service Learning."
Educational Studies 51 (1): 45–56. https://doi.org/10.1080/00131946.2014
.983637.

Sobania, Neal W., ed. 2015. *Putting the Local in Global Education: Models for
Transformative Learning Through Domestic Off-Campus Programs.* Sterling, VA:
Stylus.

Tiessen, Rebecca, and Robert Huish. 2013. "International Experiential Learning and Global Citizenship." In *Globetrotting or Global Citizenship? Perils and Potential of International Experiential Learning*, ed. Rebecca Tiessen and Robert Huish, 3–20. Toronto: University of Toronto Press.

Wallace, Tim. 2011. "Apprentice Ethnography and Service Learning Programs: Are They Compatible?" *Collaborative Anthropologies* 4 (1): 252–9. https://doi.org/10.1353/cla.2011.0002.

Ward, Martha C. 1999. "Managing Student Culture and Culture Shock: A Case from European Tirol." *Anthropology & Education Quarterly* 30 (2): 228–37. https://doi.org/10.1525/aeq.1999.30.2.228.

SECTION ONE

The Processes of Learning Out There

Authors in this section examine the processes of field-based learning from the perspectives and practices of diverse disciplines, from geography and urban studies to earth sciences to history (the latter a discipline with little tradition of field-based study). The field experiences described are also diverse, lasting from ten days to three months. Each author considers learning "out there" in a dual sense – learning well away from the university classroom and learning that takes students out of their comfort zone and aims at some measure of transformation. The learning described in some chapters is that of the instructor as well as of the student.

Geographer Kacy McKinney examines a specific kind of field-learning experience, an intensive and relatively long (three-month) immersion in a single location. This approach, she argues, is particularly effective in helping students to move towards thinking relationally. For McKinney relational thinking is rooted in feminist critical approaches, which foster acceptance of one's limits, the shared vulnerability of group learning, and a deepening appreciation of one's particular privilege and social location in relation to the field. Drawing on student reflections collected five years after their experiences in India, McKinney shows how students learned to see differently, moving from adherence to preconceptions and standard binaries to a more critical analysis of the shared human condition.

Aaron Williams recounts the experience of University of Calgary field programs in physical and human geography, urban studies, archaeology, and earth sciences. Once entirely separate from one another, the programs have been enriched by integrating disciplinary approaches and field experiences under broad themes, such as sustainability or natural hazards, and connecting these to a physical context such as a river. Case studies illustrating integration of disciplinary approaches and teaching show how students are guided to move beyond the siloing of information towards a more nuanced and comprehensive analysis of the subject of study.

Dialogue is central to the chapter by historian Nakanyike Musisi, who describes her efforts to bring the positive elements of two short-term field programs – one examining causes of conflict in African states, the other using individual life histories to explore historical questions – back to campus. While in the field, deliberate and frequent shifting of roles, including by the instructor, helps to build community and prevent interpersonal conflict from ossifying. Musisi explores ways to bring that intimacy and dialogic practice back to the university classroom. She works to replicate a "we're all in this together" community of learners where questions take priority over answers, and ambiguity over certainty.

As geographers and planners, Heather Castleden, Kiley Daley, Vanessa Morgan, and Paul Sylvestre consider how to bring about meaningful engagement with Indigenous knowledges among non-Indigenous (or settler) students preparing for careers in resource management. They describe a process of guided transformation in which students spend time interacting with Indigenous knowledge holders, and then articulate their experiences – including the first steps of unlearning – through digital storytelling. Transformation, the authors argue, requires disorientation and coming to know what one does not know; openness to new learning through new relationships; and a journey through vulnerability towards pride in expanded understanding. Digital storytelling provides a space where such transformation can begin.

Student vignettes in this section reflect in diverse ways on the opening that may happen in field settings. Sam Kerr reminds us of how the past remains written on the present, in plain sight if we open our eyes. Emily Halvorsen speaks to the value of being open to the unexpected moment, embracing being "lost and confused." Rob Cook surprised himself with his openness to a classroom without hierarchy and few of the usual rules in place. These vignettes indicate the liberatory potential of some of the smallest moments in the field.

WHERE THE PAST AND PRESENT INTERSECT ...

As we walked across the sun-dappled cobblestones of the Company's Gardens in Cape Town, we passed office workers on lunch breaks and other groups of travellers taking in the pleasant shade of the afternoon. As our group made its way to the end of the garden, we passed an opening in which an imperious-looking bronze statue was situated surrounded by the ferns and palm trees of the garden. My fellow students continued along the path, while I looked up at the statue atop the stone plinth. The figure's left hand pointed beyond the gardens while his right hand held a tarnished hat against his side. People mingled casually along the adjacent paths, while our group made its way past the statue and to the end of the gardens. At that moment, the statue seemed to bring into sharp relief the influence that the past can have on the present and how our present circumstances are influenced not only by the immediacy of our current predicament but also by larger forces that are outside of our control. This statue, located in the middle of the square, also seemed to me to be a reminder of how a society can be of two minds about its history.

I learned many things while studying off-campus, but this moment in the Company's Gardens in Cape Town was particularly memorable as a history student. That morning we had returned from a workshop hosted by several community activists at the University of Cape Town, and this statue seemed to stand in sharp contrast to many of the ideas that we had discussed hours before. In fact, the very ethos of the statue seemed to stand in sharp contrast to the values of the institutions that were adjacent to the park – where the statue's honouree had been elected as prime minister. The statue's designer had captured a set of ideas that seem at best an anachronism, and at worst an anathema, in the present moment. These ideas, however, seemed to shed light on the history of the place I was in, where it had been, and where it could go. For me, studying the past often means encountering and questioning ideas that people in the past have propounded and relating those to where we are today. While passing by this larger-than-life statue in the shaded gardens of Cape Town, the influence of our collective history on the present moment became particularly clear.

Sam Kerr
Participant, University of Victoria's Colonial
Legacies Field School in South Africa

1

"You Cannot Avoid All of This Past, Present, and Future When It's Everywhere Around You": Reflecting Relational Thinking in Field Study Experiences

KACY MCKINNEY

Introduction

The single story creates stereotypes, and the problem with stereotypes is not that they are untrue, but that they are incomplete. They make one story become the only story.

– Chimamanda Ngozi Adichie (2009)

As a feminist critical geographer, one of my central pedagogical aims is that students learn not to rely on a single story to explain the world around them, but instead that they work to trace complex connections across time, space, and place. I am interested in exposing students to the situated nature of knowledge, training them to seek anti-essentialist approaches, and inspiring in them a dedication to exploring multiple perspectives on any issue. But how do we teach a way of *seeing* the world and our place within it?

Field study in a single location offers a unique means through which to achieve this aim. Studying intensively away from the traditional class-room, students can learn to shift from emphasizing differences, or engaging in processes of "othering" (Said 1978), to thinking relationally. For geographers, relational thinking is an approach to making sense of difference and similarity (Jackson 2006), in which we explore how we are connected rather than uncritically defaulting to distancing binaries. Indeed, when students learn to think relationally, they no longer accept simple explanations. Relational thinking is a form of critical pedagogical praxis. As students are guided through new and shared experiences in unfamiliar places, they work to give locations meaning and to build understandings of them (Schmidt 2011). Students engage in the processes of making sense of place and of shaping their identities in relation to new places (Tuan 1977), and many come to understand a global sense of place (Massey 1994). Following Nairn (2005), I want to emphasize that it is not through direct field experience *alone*, but rather through field learning experiences that are based on clear critical and feminist pedagogical motivations and intentions, that engagement in field learning can help students to develop the ability to think relationally.

In an effort to understand the specific mechanisms through which students come to think in this way, I reached out to the participants of two intensive three-month-long undergraduate programs on the topic of international development and environmental change, which I co-directed in rural India in 2011 and 2012. In this chapter, I am drawing on the reflections of these students, who completed their respective programs more than five years ago. These reflections at once offer a rich set of perspectives on what these experiences have come to mean over time, and demonstrate the development of relational thinking. I present these as illustrative examples, rather than as a representative or generalizable study. In the tradition of feminist qualitative research, and along with the other authors in this volume, I seek to contribute grounded experiences to a growing body of research on pedagogy and field study.

The teaching philosophy that underpinned this field study program draws from feminist critical pedagogy in its intentional focus on critical social analysis; its insistence on the situated and socially constructed nature of knowledge; its emphasis on the role of experience and emotion in knowledge formation; and its commitment to anti-essentialist and intersectional practices and dialogue in the examination of structures of oppression. In this approach, I am particularly inspired by Boler (1999), Haraway (1988), hooks (1994), Peake and Kobayashi (2002), Rabinowitz (2002), and Shrewsbury (1993). Critical feminist pedagogies, as Maxine Greene puts it, "demand critical examination of what lies below the surface. They demand confrontations with discontinuities, particularities, and the narratives that embody actual life stories. At once, they require renewed attentiveness to the construction of knowledge and the life of meaning" (2013, x). Critical pedagogy views education as the development of critical consciousness and as social transformation, education in which power and oppression are addressed in and through teaching and learning. With its origins in the influential works of John Dewey (1916),

Michel Foucault (1980), Paulo Freire (1971), Antonio Gramsci (1971), Maxine Greene (1988), and others, critical pedagogy has been strengthened where it has been responsive to feminist critiques, and in critical feminist pedagogy I see the foundations for relational thinking. In the reflections presented here, students look back on experiences within an intentionally critical feminist field study program.

In the analysis of these reflections, three interrelated themes, which might also be seen as a progression, emerged. These reveal some of the ways that students learn to think relationally, and they shed light on benefits and limitations of field study design, teaching, and assessment. The first theme relates to the processes by which students come to accept the limits to their knowledge and the complexity of interwoven economic, social, political, and environmental processes across time and space. The second theme relates to the power of learning as a group in new and challenging circumstances, and points to how shared vulnerability can deepen student learning. Finally, and perhaps most importantly, the third theme relates to how students develop a particular way of seeing the world as connected and how this is done through the recognition of their privilege and positionality. This evolution of world view is one of the most exciting possibilities posed by field study as a pedagogical tool aimed at learning to see the world and the self differently.

In the next section I offer background on this program, before turning to a discussion of the three themes. In the final section I address two sets of practical lessons that also emerged: first, student reflections elucidated significant distinctions between classroom and field learning, which can be instructive as we argue the case for continued investment in well-defined off-campus study as a component of the overall undergraduate experience; and second, student reflections speak to the importance of the time needed to understand and articulate the lasting impacts of intensive field study participation.

The Environment and Development in the Indian Himalayas Program

In the fall of 2011 and the summer of 2012 I co-directed a three-month study abroad program in rural India, together with my colleague, Keith Goyden. The program had been running for several years and has continued in the years since. In the two programs we had a total of twenty-six participants, just two of them male. The program is offered through the University of Washington's Jackson School of International Studies in the United States and draws undergraduate students from a range of disciplines. The emphasis of the program on environment and development draws interest particularly from students in political science, geography, anthropology, international studies, and public policy. The program is composed of two courses that each meet twice weekly as seminars, an internship or fieldwork component in conjunction with a local non-governmental organization that runs throughout the program, and a one- to two-week stay with a local host family. Other than during the stay with a family, the students, the co-director, and I lived together, ate together, and had opportunities for conversation throughout the course of every day.

The two courses are "Political Economy of Indian Development" and "Gender, Work, and the Environment." Course readings for both of these are drawn from across the critical social sciences and specifically focus on postcolonial studies, feminist political ecology, development studies, political economy, and environmental history. A significant portion of the literature comes from South Asian writers, and much of it is focused on the specific area of India where the program is located, in the state of Uttarakhand. Students are paired with an Indian organization based locally that addresses wide-ranging issues from women's health to agricultural innovation.

During the program, I observed that many students at first approached India with a focus on all that was strange; they viewed their surroundings through a lens of what stood apart from experiences back home. They often sought to compare and to list all that was striking, different, exotic, and bewildering. They sometimes became consumed by what did not appear to make sense, creating a distinct barrier to learning. After some time, however, they began to see how they were connected to their surroundings and to take note of fundamental issues of the human condition. Through everyday encounters many students began to recognize some of the patterns they were reading about, and we took the time to discuss, observe, engage, listen, and question what they saw and read. Certainly, students did not all progress through this change at the same pace, but they helped each other along throughout the program. Even as I observed changes in these students, I was left wondering both how the changes had come about and what these experiences would mean in time.

For this chapter, I reached out to all of the participating students and requested their reflections on four key questions. These questions were meant to spark reflection, and students were welcomed to write as much or as little as they felt inclined to do and to stay within the bounds of the questions or extend beyond them. I framed the questions as a means of reflection on their experiences and the longer-term impacts of the program on them rather than as an evaluation of the program itself. I received responses from half of the students, equal numbers from each group, and only one of the two male participants.

Questions Asked of Former Program Participants

1 Do you feel that the experience changed you? If so, in what way(s)?
2 On a related note: would you say that it was transformative (is there another word that you would use to describe the experience's impact on you)? Please explain.
3 Is there a moment(s), an interaction, a relationship, or some aspect of the program that had a particularly lasting impact on you?
4 What stands out to you in reflecting on the difference between taking credit courses in the context of the program versus back home?

Students looking back across five years have varying recollections of their time in India, of how they learned, and of what they experienced. I offer these reflections as illustrations of learning processes and spaces of possibility for learning in the course of field study. My aim is to suggest patterns in how a particular field study program impacted student thinking.

Theme 1: "India Struck Down My Confidence by Continually Proving Me Wrong"

I applied for this program with so much confidence. But two seconds into landing in Delhi, all of that confidence eroded. India seems to do that though. India defies every expectation … After the program, I feel as though I approached things with a new attitude. I don't understand the full story. I do not understand the human emotion and reality that drives facets of the story. It is my job to approach everything with an inquisitive and passionate mind, but, more importantly, with humility. (VE, pers. comm.)

Over the course of the two programs, I observed a fragile confidence with which many students initially clung to single stories about India and to a faith in their capacity to interpret the world around them. This is a somewhat intensified version of something I have also noted in the traditional classroom environment. At the risk of feeling overwhelmed by the complexity of the world around us, we tend to hold tightly to the vocabulary of binaries and simplified geographical reference points such as "First" and "Third" worlds and North/South dichotomies, for example. In the course of field study in rural India, students were initially frustrated when faced directly with complexities such as this. But over time, many of them arrived at a different place, beginning to accept, and even expect, the unexpected. For example, one student reflected:

> I became incredibly comfortable with the unexpected in a way I had never experienced in the past. I loved the unexpectedness of each day and not knowing what I was going to experience or encounter. (KD, pers. comm.)

As I read through the reflections from former students, I was struck by how poignant this lesson had been for many of them. Growing comfortable with the vastness of historical, cultural, social, and political context surrounding us, for example, and expecting the incredible multiplicity of perspectives on any issue is not only productive, but also necessary for feminist, critical, and justice-oriented thinking. This recognition means that we are responsible for constantly questioning, listening, and expecting that there is more to any story. This process is not an easy one to go through, for it involves a sort of loss of the ground under one's feet.

As the student VE notes in the quote opening this section, while at first this sudden erosion of confidence can come as an overwhelming surprise, it can turn into a source of constant curiosity and seeking. As you begin to see the connections between processes and people across time and space and let go of the impulse to need to understand everything, the world becomes an exciting, unknowable puzzle. This, I believe, can inspire both lifelong learning and the drive for social justice. The same student continued:

> The India program left me every day, but especially at the end of it, feeling like I actually did not know anything. I probably learned more during my time in India than during other quarters, but in a way that left me sure that I really did not understand anything. I constantly wanted more information, more sides to the story ... When you start looking at the connected strings of how this [management of forest resources] has changed the lives of the people today: Male outmigration towards the cities, education, women's increased household responsibilities because of the outmigration, the

shift to cash-crops ... You cannot avoid all of this past, present, and future when it's everywhere around you. (VE, pers. comm.)

Another student echoes this sentiment and suggests that the combination of academic coursework, on the one hand, and learning through everyday encounters, building relationships with local people, and participating in NGO work, on the other, is what helped to build this changed awareness and recognition of the limits to her knowledge and understanding. She states:

Even five years after completing this study abroad program, I continue to reflect on the ways in which it affected me [personally], and my academic trajectory. One way I feel changed is in my awareness of the nuance of place. This awareness developed out of the program's academic and experiential approaches to learning ... both pieces complimented [sic] the other and complicated my developing understanding ... I remember leaving India with more questions as well as more of an awareness of my own ignorance. (ESP, pers. comm.)

While at first it can be extremely frustrating to feel the weight of how much we do not know and cannot understand, the growth in awareness and acceptance and even expectation of complexity that these reflections demonstrate is a critical piece of learning to think relationally. Students experienced a change in perspective – although not all did so in the same order, at the same pace, or in a smooth manner.

Theme 2: "[It Gave Us] an Opportunity to Get Comfortable in Our Shared Vulnerability"

The experience was absolutely transformative for me. I think it brought me out of my comfort zone in more ways than one and forced me to rise to many challenges that I would not have had to otherwise. (AB, pers. comm.)

From Day 1 of study abroad in rural India, students expressed feelings of frustration, concern, of being wholly overwhelmed, and of losing confidence. I struggled to comfort them, offering stories of my own awkward and sometimes painful early days of studying abroad in Brazil and India. In an effort to address these feelings in students, my critical feminist colleague, Keith Goyden – who has vastly more experience working with students through field study – showed Brené Brown's 2010 TedX Talk "The

Power of Vulnerability" to both groups. In the talk, she argues that personal and relational "connection is why we are here" and that we embrace vulnerability when we have the "courage to be imperfect." According to Brown, "In order for connection to happen we need to let ourselves really be seen." Brown's ideas led to important conversations early on in the two programs, and they remained salient to participants five years later. One student wrote:

> [The program gave us] an opportunity to get comfortable in our shared vulnerability. I had never actively thought about the power of vulnerability before going to India and finding myself in situations I had never imagined, and finding myself there with my wonderful cohort gave us a venue to feel vulnerable together. (KD, pers. comm.)

For many students, studying together, as a group, played a central role in the lasting impacts of the program on them both academically and personally. Other students also dove into the dynamics of shared experiences of vulnerability and intimacy during the program:

> There are inherent challenges that come with living with peers in somewhat of an isolated environment without a lot of freedom. I wouldn't say I always handled the challenges with grace, but being forced to deal with social issues among my peers helped me to understand what it means to be part of a group and how to constructively handle challenging situations. I remain close with many of my peers, and in part I believe that's because all of us, in many ways, helped one another develop as people and mature. (SC, pers. comm.)

A critical piece here is the importance of shared experience for deep and lasting learning. Students emphasized that together they were able to feel secure in the face of challenge and discomfort, and they could support one another in those vulnerable moments. This aspect of field study – extended time spent as a group – is a unique pedagogical tool for building relational thinking. As students recognize their shared vulnerability, they meet challenges as a learning community, and in doing so discover new ways to connect with the world around them. Two students spoke to this point:

> My experience studying abroad in India was incredibly transformative, and the friendships I made through my program were one of the most valuable

things I took away from the program. Being so immersed in my studies with a close-knit group of students was so valuable personally and educationally. We learned from each other through sharing insight and experiences and were able to connect with each other on such a deep level. (KD, pers. comm.)

The intensely close connection that the study abroad fostered made us learn better and provided us a network of support when things got hard or confusing. We processed a lot of the new world around us as a group, helping to give context to the experience. (KW, pers. comm.)

This shared vulnerability involves a process of developing intimacy, coming to trust one another, and learning lessons that have made them stronger individuals. Their reflections demonstrate a great sense of lasting connection and the powerful nature of their shared learning and experiences for developing who they are today. Another student wrote:

I think that being in such an intimate space with my cohort created opportunities for learning from the other students in a way I rarely had at UW. We had emotional conversations about privilege, emotional responses, inequality and power, and these charged yet academic conversations taking place in a space very new and different to all of us propelled us into a level of intimacy I rarely found with other students at UW.

One afternoon, I was walking with a friend from the cohort, and he asked me a question I haven't forgotten. He said, "[AD], are you proud of yourself? Not will you be proud of yourself, but are you proud of yourself right now, in this moment?" And that is something I think about every day. We should be what we want to see in the world. (AD, pers. comm.)

These are not easy questions, and they are not straightforward conversations. Just as students were losing confidence in their preconceptions about India, and finding themselves uneasy and uncertain about many things, they were creating space within themselves to grow and to accept the limits of their knowledge. They helped each other grow comfortable in sharing the stumbling blocks and defeats alongside the growth and maturation inherent in new and challenging situations. Brené Brown's work helped the students, as well as my colleague and me, put words to this complex and rewarding experience of shared vulnerability.

Theme 3: "The Experience Shifted My Understanding of the World and [My] Privilege in It"

Over time, students became increasingly able to reflect on their place in the world, their previous experiences and their ideas, and they began to demonstrate an increased capacity for tracing connections across time and space. One student wrote:

> I remember a conversation I had with one of the families I met with in India, in which the only question they wanted answered was what crop my family grew back home. Such a simple question, and yet it took me totally by surprise. I felt silly, sitting there with my notebook and Western-minded worldview, with an outlook so limited by grocery stores and ease-of-access. It made me stop to think about how much my lifestyle limited my ability to connect with the things I rely on to survive, such as my food, and with the people who provide it to me. (AS, pers. comm.)

In her reflection, this student hit upon a critical shift for students, moments in which they begin to see connections that had previously been invisible to them. The networks and relationships that underpin food production and distribution is a common example, but there are countless others. Once students in the program began to see threads of connection that had been hidden to them, they became curious about what other connections they may have missed. This curiosity is at the heart of relational thinking.

The wide range of new experiences led to discussions and reflections on privilege, as students wondered what they may have missed and what they may have been sheltered from. For many students, the program was filled with firsts related to a heightened awareness of their own positionality. One student wrote:

> Because we were visiting the space(s) we were studying, it was also the first time I became aware of the role of the individual in development, government, workforce, etc. in a way that would not have been possible in the classroom. It was the first time I think I really had to confront and assess my privilege and think about my role in the global community. (JK, pers. comm.)

This same student continued:

> I was a transformed human ... It was the start of feeling like I needed to contribute to the world, as the program provided direct exposure to the

cultural and social contexts we were studying, I finally understood that inaction, complacency, and complicit support contributed just as much as any action to constructing systems of injustice. Throughout the program, I remember that my main focus was on determining and discovering how to quantify "Sustainable Development," and this was really an exercise in understanding that constructs are flexible, and whoever utilizes the term, is also implicitly defining it from their contextual perspective. (JK, pers. comm.)

JK articulated her focus on exploring the idea of sustainability from Day 1 of the program. She asked most people we met: "What does sustainability mean to you?" The response from one woman left a lasting impression on all of those present. She said:

> It means knowing the difference between what you *want* and what you *need*"
> (original emphasis).

JK, and other students, did much reflecting on that definition of sustainability and how it connected to their own lives.

Through conversations and interactions over a period of three months, students heard about violence, inequality, courage, and hope. They listened to different approaches to everything from consumption of resources to love and relationships. They found, in the stories and reflections of the people we met, rich connections to life back home, and they were forced to consider their own positioning and privilege in the world. Another student reflected on the impact of stories:

> Studying abroad in Uttarakhand absolutely changed me ... I heard stories about how hard some girls I met and shared meals with had to fight just to be able to go to school, and spoke to women about their *decades-long* work to be able to purchase a *cow*. These stark reminders of inequality both inspired me to never take my privilege for granted, and also put me in awe of the strength of the women I met. (KD, pers. comm., original emphasis)

Listening to the stories of women of all ages and from varying economic backgrounds and communities, who, in the face of seemingly insurmountable challenges, continued to struggle and support each other, impacted these students in lasting ways. And even without knowing the extent to which students continue to reflect on and understand their privilege, this awareness is a key point of reflection in their responses, and something they connect directly with field study. Another student reflected on these changes:

> My entire world view changed ... I went into the trip believing that I alone could change and shape the world, both in the environmental field as well as the gender equality field. India was a wake-up call and really opened my eyes to the fact that a foreign, white female could not, and maybe should not, shape the gender dynamics and environmental issues facing the country ... The trip changed my perspective from a naive, almost childish view to a more practical and focused view. I am pursuing water law because I know how large of an issue water is both locally and globally. (EH, pers. comm.)

These shifts in how students see the world and their place within it are made possible through everyday interactions, listening, observing, being vulnerable, accepting the limits to their knowledge and experiences, and being in place intensively studying and living abroad. Such changes in students are perhaps the most powerful aspect of a field study experience, and they represent a means through which field study pedagogy can be used to push students beyond binaries and single stories.

These three themes draw out important aspects about field study and relational thinking. Through the breaking down of preconceptions and opening up to new world views, the building of shared vulnerability and togetherness, and critical reflection on privilege and positioning, students can learn to think relationally. In the next two sections, I turn to how students differentiate studying at home versus through field study and to the question of the time necessary for reflection on the personal and academic impacts of this form of study.

On the Difference Between Classroom
Learning and Field Learning

In their responses to my question on the differences between classroom learning and field learning, students reflected on what makes the latter distinctive, even necessary, for building relational understandings of the world. Students described learning through field study as collaborative, experiential, necessary. For one student, the rigour of the program, the small size of the cohort, and the opportunity to develop relationships with her teachers was fundamental to her learning at the time, and it led her to transfer to a small liberal arts college on completion of the program. Several students emphasized these connections between ways of learning. Deborah Curran (this volume) refers to the constant and reflective learning in place made possible through field study as "embeddedness." One student described it in this way:

> Taking credit courses in the context of the program solidifies the lessons of the courses in a way that no normal credit course back home ever could. There is no going home at night and forgetting what the professor lectured on in class that day because that lecture has just become your lived experience. It's a more extreme form of experiential learning: the subject matter has become what you are surrounded by on a daily basis and after that it becomes impossible to forget. (KW, pers. comm.)

On a similar note, another student wrote about the significance of the unique schedule and mechanisms of learning in an intensive field study program:

> [It] was vastly different than my experience taking classes at the university. Unlike at home where learning and studying was confined only to the hours in the classroom or library, studying abroad gave me the opportunity to be constantly learning, observing, and gaining valuable insight from the people and culture I was immersed in. While my study abroad program still had set times for classes and studying, the learning was so much deeper and holistic than any of my classes at the university had been. I was able to learn through experiences, adventures, and through the relationships I had built. (KD, pers. comm.)

Another student also wrote of the significance of encounters and immersion for learning:

> [What stands out to me is the] direct exposure to the contexts you are studying and the role of community and support in education and personal growth. Both of these aspects are lacking in a traditional lecture-style university course. The fact that we were living in India and interacting with each other (peers, instructors, community members) 24/7 for an extended period of time, made experiential learning, and reinforcing the learning we were doing, accessible. (JK, pers. comm.)

These reflections demonstrate an integration of relationality into participants' world views and make a compelling case for field study programs as a means to engage in the critical pedagogical praxis of relational thinking. The emphases of these students on the immersion of field study, the blurred boundaries of learning and daily life, and the unique space created by the shifted nature of the classroom, speak to how field study lends itself to unique ways of thinking and learning. One student wrote:

> Ultimately, it is impossible to *teach* a culture – it must be experienced. And, because our society is built upon relationships with people from other viewpoints and ways of life, it is crucial that we are able to understand the framework through which our neighbors see the world. Study abroad facilitates this in a way impossible to do through a campus-based class or experience. It is in the context of our politically divisive, emotionally charged, hate-filled world that study abroad is so crucial to our survival and progress. (AS, pers. comm., original emphasis)

Another student emphasized the process of reading academic work in place – of seeing the patterns and relationships, the social, environmental, and political processes playing out in real time – and of the lasting impact this makes. She wrote:

> Taking credit courses in the context of the program offers the invaluable experience of living the material of which you are learning. At home you can read and learn all about gender inequality around the globe and how it manifests itself differently depending on which country you are in. But being in the same region that is discussed in the readings is absolutely indescribable. You feel the same passions that the writers feel. You can see the trees chopped in precisely the same way that you read about, and what's more is you understand why they are chopped in that manner. Every time you see a water buffalo you think of the woman who is caring for that buffalo, in addition to her house, her crops, her husband, her children, and her husband's family. When at home, you read about those same gender dynamics and brush it off as a fact of the world. When you live in the classroom material, experiencing it every day, it takes on a new and deeper meaning. It resonates and sticks with you more than the material you learn at home. (EH, pers. comm.)

Learning about gender inequality, and other forms of difference, in a way that draws on and reinforces critical feminist empirical and theoretical scholarship opens up the possibility of recognizing and understanding contexts of inequity and of seeing openings for change. Another student echoed this difference of reading academic work in place and the deep impression it makes:

> There is a world of difference between reading an academic paper about the long-term and hugely painful injuries that many Indian women sustain from carrying heavy loads in their daily chores and seeing a tiny woman smile as she walks down the road with 50+ pounds of apples or pears or potatoes on a crate on her head ... It's all well and good to postulate the benefits of local development work in rural India, but a whole different thing to see the benefits yourself. (KW, pers. comm.)

Comparing studying at the university and abroad, another student raised the question of evaluation, and how the emphasis on grades was diminished for her through the act of learning in place, writing:

> In India, I rarely thought about assessment, because I was so swept up in processing both the academic work and the personal growth I was going through. I think ultimately that made me produce some really excellent work, on a very short time scale. I would also like to say that the relationships I built with my professors ... were so comfortable and productive and safe that I was not at all concerned with assessment ... And a brief side-note, because of the level to which we were "unplugged," ... I somehow felt less pressure, less distraction, and more of a creative drive to write and learn. (AD, pers. comm.)

Many students felt that the internship or fieldwork component – the part of the program in which students worked directly with NGO staff, interviewed local women and men, and observed development work in the area – was one of the most significant experiences of their undergraduate careers. One student described it in this way:

> The experience of working with nonprofit staff, interacting with the community, and being able to shape and guide my own project was something that you can't replicate in the classroom ... The skill set gained from working with people of another culture, adapting to challenges in the field, and remaining accountable to the success of a project that would be used by an organization are immensely valuable and applicable to the "real world," and are exactly the skills that you can't learn in a traditional classroom. (SC, pers. comm.)

Framing field study as a particularly effective means through which to learn about both different cultures and how we are all connected, these students highlight immersive experience, interaction, connection, and relationships as critical components. This poses an interesting pedagogical question: what elements of this form of learning to think relationally can, in fact, be brought back into the traditional classroom? In my regular university courses, I am constantly seeking ways to draw in direct experience and modes of connection to people and places outside of the classroom – working to bring the lessons of field learning into the classroom (see Musisi, this volume). At the same time, the embeddedness referred to by Curran (this volume) makes field study uniquely positioned to impact student understandings of relationality in lasting ways.

On the Significance of the Time Needed to Reflect

I think about my first time in India frequently. It may not be an explicit memory of my time there, but there were so many lessons, feelings, relationships, confusions, perspectives, concerns, and thoughts that remain with me even five years later. (VE, pers. comm.)

A striking element within these reflections was the emphasis on the time it has taken to process the complex experiences, lessons, and meanings of the intensive field learning experience. This, of course, presents a challenge for program assessment and evaluation (see Owens and Sotoudehnia, this volume). We gave students evaluations during the final week of the program – a week in which they were writing final papers, presenting on research, saying goodbyes to new friends and places, and organizing their plans for further travel or a return to commitments back home. Certainly, students were able to articulate parts of the program that were significant for them, and they spoke at the time of both wonderful experiences and missed opportunities. But they also expressed the difficulty of knowing what the program meant to them, how they might draw on it in the future, and what the lasting impact of it would be.

Students demonstrated a richer understanding of social, political, economic, and environmental history and context in India – they gained much from the coursework and the internship and research experiences that could be measured and evaluated at the end of the program. But there were also signs that there would be lasting impacts on them individually, which we could not yet see or be sure of at the time. In this section, I draw on student reflections of the passage of time and the lasting impact of this particular program. In doing so, I am also drawing attention to the need for longitudinal study of the impacts of field study on participants. Longitudinal studies often pose challenges on account of the logistics necessary to keep in touch with generations of students, as well as through the fading and shifting of memories over time. Still, these reflections begin to give us a sense of what field learning can come to mean for some students. As with any pedagogical approach, we must recognize that countless factors are at play and that students learn in very different ways. At the same time, this should not keep us from observing the benefits to many students of their off-campus learning experiences.

Several of these students articulated that, five years on, they can see how the program impacted who they became professionally and personally. One student wrote:

> I am still surprised at how much I am able to draw from my experiences during the program and apply them to my work, and how much small, at the time seemingly insignificant, experiences still influence and inspire me as I move forward. I would say the program was life changing in the sense that it was the beginning of what I've chosen to pursue. Beyond academics, the program changed me as a person … I'm so grateful to have been offered the opportunity to meet and live with families that I did. Those experiences to this day help shape how I approach culture and social differences as an adult, whether that is in my own city or working abroad in a town or village I've never seen before. (SC, pers. comm.)

Another student wrote:

> Traveling is always a moving and eye-opening experience. My experiences with other cultures, languages, and worldviews have shaped me into the person I am today, and my time in India contributed greatly to that. Now, as I have begun my career as a Pediatric Nurse, I reflect often on my time in India. Without it, I would be less aware of my own shortcomings, and far less eager to provide culturally sensitive and appropriate care to my patients. (AS, pers. comm.)

Former students emphasize that they worked not only to understand the meanings of the experiences for themselves, but also how to share these, how to talk about them, which presents another reason for us as program directors to check back in with students over time. For example:

> My entire worldview was transformed for the better. I was exposed to a completely different way of living in a completely different environment. I was pushed outside of my comfort zone, which is the best way to learn about yourself as well as others; and I do believe that in that zone is where transformation takes place. It took me months, even years, to learn how to describe my experience in India. At first I was so overwhelmed with the experience that I could barely find any words. Slowly, over time, I discovered that the best word for my experience was "challenging." It was challenging not only because of the change in lifestyle and learning how to live in a new and different culture, but also because of the dichotomy and juxtaposition of the emotions I felt. (EH, pers. comm.)

Another student wrote:

> When people ask me about my time studying in India, I say that it was the absolute worst experience of my life. I describe how crowded, and hot, and dirty it was. How nothing ever went according to plan, and that it was emotionally and physically exhausting ... Then I say that it was also the best experience of my life, and that I miss it every day. It was not easy to be away from home, to be without the luxuries I took for granted back in the U.S., or to witness the sexism that was pervasive in my everyday experiences. But it was those stories that made me laugh and cry so much afterwards, and now, five years later, still remain with me. (AS, pers. comm.)

Speaking to several related issues at once, another student wrote:

> It's often very difficult for me to eloquently describe what the program meant to me, in part because I think I am still in the process of discovering that for myself ... It was an experience. It was a turning point. The program has and continues to help me figure out things about myself, about what I want to do with my life and why, my values, and what it really means for me personally to find happiness. I still realize things about the program that affected me five years later, and I think that's what was really "transformative" about the program. It was something that was wonderful when it happened, but still continues to impact my life and guide me as I have new experiences. (SC.2, pers. comm.)

These student reflections show some of the ways that a program of field learning based in critical feminist approaches can change us, and that it can offer the means through which to learn to think relationally. This change comes about through a confluence of experiences and challenges from living in place together as a group, to existing in a constant state of confusion, and often – at least mild – discomfort. Through this confluence, we can learn to see connections, to recognize our positioning in the world, and to accept how much we do not know.

As co-directors, Keith and I did not seek to *teach a culture*, but rather to create a program based in what Boler (1999, 166–7) refers to as a "collective engagement in learning to see differently" (quoted in Nairn 2005, 307). We avoided some of the potential ethical, practical, and theoretical problems with using direct experience as a primary means of study about the world (Nairn 2005) by conducting the program with clear pedagogical intentions and motivations based in critical feminist practice. Students learned about the history and political economy of India, about forest resource management, and rural women's everyday lives; they learned about local impacts of international development in theory and in practice.

In our attempt to inspire relational thinking through situated knowledge and anti-essentialist approaches to geographical knowledge, this program shows how we might teach a way of seeing the world and our place within it. Through spaces of reflection and shared vulnerability, guided reading and discussion of critical and feminist social theory, and myriad forms of engagement and connection, field learning can be used to prepare students to map connections as ways of understanding. With a central emphasis on anti-essentialism worked into program pedagogy, students can learn to reject single stories and to seek complex connections across time, space, and place.

NOTES

I am grateful to my colleague and dear friend Keith Coyden and to all of the students in the 2011 and 2012 Environment and Development in the Indian Himalayas programs, especially: Emily, Erin S.P., Erin H., Kyle, Ashley, Arianna, Katherine, Kara, Victoria, Joelle, Alice, and Sarah. Thank you to Cam Owens, Deborah Curran, and Kirsten Sadeghi-Yekta for valuable insights and feedback.

REFERENCES

Adichie, Chimamanda Ngozi. 2009. "The Danger of a Single Story." TED Talk.
    London. Accessed 23 Feb. 2017. https://www.ted.com/talks/chimamanda
    _adichie_the_danger_of_a_single_story.
Boler, Megan. 1999. *Feeling Power: Emotions and Education*. New York: Routledge.
Brown, Brené. 2010. "The Power of Vulnerability." TED Talk. Houston. Accessed
    23 Feb. 2017. https://www.ted.com/talks/brene_brown_on_vulnerability.
Dewey, John. 1916. *Democracy and Education*. New York: Free Press.
Foucault, Michel. 1980. "Truth and Power." In *Power/Knowledge: Selected Interviews and
    Other Writings 1972–1977*, edited by Colin Gordon, 109–33. New York: Pantheon.
Freire, Paulo. 1971. *Pedagogy of the Oppressed*. New York: Seabury.
Gramsci, Antonio. 1971. *Selections from the Prison Notebooks*. New York: International
    Publishers.
Greene, Maxine. 1988. *The Dialects of Freedom*. New York: Teachers College.
Greene, Maxine. 2013. "Foreword." In *Feminisms and Critical Pedagogy*, ed. Carmen
    Luke and Jennifer Gore, 1–14. New York: Routledge.
Haraway, Donna. 1988. "Situated Knowledges: The Science Question in Feminism
    and the Privilege of Partial Perspective." *Feminist Studies* 14 (3): 575–99. https://
    doi.org/10.2307/3178066.
hooks, bell. 1994. *Teaching to Transgress: Education as the Practice of Freedom*. New
    York: Routledge.
Jackson, Peter. 2006. "Thinking Geographically." *Geography* 91 (3): 199–204.
Massey, Doreen. 1994. *Space, Place and Gender*. Cambridge: Polity.
Nairn, Karen. 2005. "The Problems of Utilizing 'Direct Experience' in Geography
    Education." *Journal of Geography in Higher Education* 29 (2): 293–309. https://
    doi.org/10.1080/03098260500130635.
Peake, Linda, and Audrey Kobayashi. 2002. "Policies and Practices for an
    Antiracist Geography at the Millennium." *Professional Geographer* 54 (1): 50–61.
    https://doi.org/10.1111/0033-0124.00314.
Rabinowitz, Nancy Sorkin. 2002. "Queer Theory and Feminist Pedagogy." In
    Twenty-First-Century Feminist Classrooms: Pedagogies of Identities and Difference, ed.
    Susan Sanchez-Casal and Amie A. MacDonald, 175–200. New York: Palgrave
    Macmillan. https://doi.org/10.1057/9780230107250_8.
Said, Edward. 1978. *Orientalism*. New York: Pantheon Books.
Schmidt, Sandra. 2011. "Theorizing Place: Students' Navigation of Place Outside
    the Classroom." *Journal of Curriculum Theorizing* 27 (1): 20–35.
Shrewsbury, Carol. 1993. "What Is Feminist Pedagogy?" *Women's Studies Quarterly*
    15 (3/4): 9–16.
Tuan, Yi-Fu. 1977. *Space and Place: The Perspective of Experience*. Minneapolis:
    University of Minnesota Press.

LIVING IN THE MOMENT ...

The rolling foothills of the Himalaya are brown, terraced, and shining in the afternoon sunlight, as I stand atop one and gaze over the valley. Directly in front of me stand the majestic Himalayan peaks: clear and bright and tall as clouds. I ponder the complexity of the region – the beauty with the conflict – and my thoughts stray to my current situation. I am deep in the rural hills, miles away from my homestay, which is in turn miles away from my safe-haven, Sonapani. I am a woman all alone with a male guide, who barely speaks English and is discussing work matters with a farmer who owns the land on which I stand. I am lost and confused, but completely at ease as I drink in my surroundings.

A movement captures my attention. Through a field of Cannabis plants stands a small child, smiling and giggling. I smile in return and place my palms together and say "Namaste." The small child returns the greeting and then laughs. He runs behind another plant and says "Namaste" again. I duck my head behind a different plant and return the greeting. This peek-a-boo game continues for another thirty seconds, and I end up laughing just as much as the child.

I do not know why the child initially began laughing, nor do I really care why. It was spontaneous and fun. It is so easy to lose yourself in the formalities of life that, when you are presented with the raw spontaneity of reality, it seems almost foreign. This brief moment of playing "Namaste-peek-a-boo" in a field of Cannabis on a gorgeous fall day in the foothills of the Himalaya always reminds me to appreciate each moment, and, while it is beneficial to plan ahead, one must take time to live in the moment. Among the conflict and the discomfort, still lies beauty and laughter.

Emily Halvorsen
Participant, University of Washington's Environment
and Development in the Indian Himalayas

2

An Integrative, Thematic Approach to International Field Study Programs

AARON WILLIAMS

Introduction

In this chapter I share insights from my experience running geography, urban studies, and earth science field programs through the University of Calgary over the past nineteen years. In particular, I highlight the value of integrative, thematic field schools, by which I mean travel study programs focused around a particular theme, such as sustainability or natural hazards, that encourages the productive integration of approaches to inquiry from the physical and social sciences. Such programs can inspire students to recognize complex interrelationships that constitute the socio-natural world, propelling them beyond conventional and unhelpful binaries of culture/nature and theory/practice. In what follows, I explain the motivation for my colleagues and myself to develop integrative, thematic field programs, detail specific elements and pedagogical tools we have employed, and provide case examples that highlight the value of an integrative approach to learning out there.

Motivations for Integrative, Thematic Field Study

The rationale for learning out there is well developed in this book. Field study programs provide students with real-world experiences enabling them to better grapple with the complexity of challenges facing the planet. Employers increasingly value those who can integrate critical thinking with applied skills. International programs provide a valuable global component to a student's undergraduate education (see Peifer and Meyer-Lee, this volume). In a globalized, interconnected world, there is a demand for university graduates who can see the big picture and respond to real-world problems of the twenty-first century that rarely conform to national boundaries (Cantor 1997; Freidman and Myers 2005; Vaira 2004; Hains and Smith 2012).

In our own field study programs at the University of Calgary, we have found that the observation skills and methods for interpreting and understanding places and environments that students develop in field study often translate into greater success in students' post-field study academic work. Following their participation, many of our students demonstrate a more comprehensive understanding of the subject matter, which is conveyed in the quality of their academic submissions, their dedication to academic rigour, their enthusiasm for and interest in their studies, and their overall satisfaction with the entire program. Particularly noteworthy is the observed ability of many students to apply field learning to subsequent theory-based courses taken on campus following their field study experience. The insight and perspective students gain from a field study program often show up in the application of examples and experiences from field study to their coursework. Furthermore, we often witness students applying what they have learned beyond the classroom through activism, the formation of clubs or organizations, or through the student's current and future employment.

The increasing recognition of the importance of internationalizing higher education is further motivation for developing such programs in geography, urban studies, and earth science. At Canadian universities, 96% of institutions see internationalization as a priority within their strategic planning, and 80% see internationalization as one of their top five planning priorities (Universities Canada 2014). Fully 97% of Canadian universities provide some sort of international experience, with 70% offering international field programs (Universities Canada 2014). At the University of Calgary, where I run field study programs, it is an ambitious institutional goal as part of our strategic "Eyes High Strategy" for 50% of our undergraduate students to have an international experience as part of their degree (University of Calgary 2013). However, in spite of the emphasis on increasing the number of students participating in an international experience, we are currently falling well below the 50% mark, with only 1207 of 22,813 (5.3%) undergraduate students at the University of Calgary participating in some sort of international experience in 2015–16 (University of Calgary 2016). Although these figures would increase if considering a four- or five-year undergraduate degree, they are still well below the 50% goal.

Faculty members at the University of Calgary have been directly motivated to develop integrative, thematic field study programs by the recognition of the shortcomings of our earlier offerings. As introduced above,
we define a thematic field program as one that centres around a cognitive theme such as sustainability or globalization, connected to a specific
physical feature, such as a river or a region. "Integrative" refers to the
integration of sub-disciplinary approaches (i.e., human, physical, and
urban geography) and/or between disciplines (in our case, geography,
earth science, and archaeology) in the delivery of the field school. Thus,
an integrative, thematic field school differs from other programs that are
more general, fixed in one location, or centred around a single discipline
of study. For example, "Sustainability and Environmental Management
in Mekong Sub-region" would entail a thematic focus on sustainability
and environmental management within the defined area of the Mekong
sub-region of Southeast Asia.

Our motivation for such an approach partially arises out of our earlier
experience running (for credit) field school programs, from the 1980s
through to 2008. Such programs were delivered as rigidly separated
courses in the human geography and physical geography of a particular
region. Few connections were made between the course material and subject matter presented in each course, and there was limited consultation
and collaboration between the two geography faculty members who ran
the field school. Often approaches to problems and subject matter only
scratched the surface because both students and faculty viewed the issues
addressed in a compartmentalized way. While faculty got along well enough
and students enjoyed the programs (as indicated in course evaluations),
opportunities for synergies between faculty and course material were missing. The program reinforced the fiction that there are physical and social
worlds to be studied separately. It became difficult for our department
to justify offering these resource-intensive and costly programs without
being able to demonstrate that they were in line with our departmental
and institutional educational priorities – priorities that centred around an
integrative approach to unpacking issues of sustainability, globalization,
and environmental management. Our field program undermined departmental priorities and reinforced the long-standing great divide within the
discipline between human (social and cultural) and physical geography. It
is a fragmentation that too often characterizes classroom learning as well.

Although this potential fragmentation is highlighted in geography, it
extends well beyond. The separation of the physical and human worlds
has long been a central tenet of modern (European) thinking. As a result

of the way we are socialized to view the world, we often see our human selves as separate from our environment in both analysis of place and in our daily lives (Mark 2012). This problem extends into how students are able to use the knowledge and the concepts they have learned in classrooms and apply this knowledge to real-world situations, including employment. The dangers of fragmented thinking are, in part, the motivation for political ecology as an approach to inquiry. As Robbins (2012) elaborates, environmental issues cannot be understood simply through the physical sciences divorced from the political-cultural contexts within which they are set. Yet, as McDonald and Patterson (2007) point out, the analyses generated within urban planning theory and related disciplines are often limited by a lack of attention to important ecological or biophysical dimensions. Political ecology seeks to integrate attention to social and ecological processes. This approach motivates our program. We see field study focused on a complex and multifaceted theme, such as sustainability, as an opportunity for comprehensive, multidisciplinary, and inquiry-based learning that traces interconnections in real-world contexts using multiple methods.

In response to what we saw as the deficiencies of our fragmented field-study program, in 2007 we began planning our first integrative, thematic program in Europe under the title "Sustainability and the City: Learning from Europe." Our experiences from this program have set us on a trajectory of increasing integration of course material and collaboration between our human, urban, and physical geography faculty in the offering of these programs. Measured by course evaluations and by departmental and institutional feedback, the changes we made in the delivery of our programs have been very well received. Our offering of field study programs has seen significantly more buy-in from faculty and greater support from our institution.

The success of the programs has not been easily achieved. Effective integration and collaboration is by no means straightforward. We turn now to look at the many factors, ongoing challenges, and experiential-learning tools and methods that we have developed to offer promising integrative, thematic field study programs.

Tools to Facilitate Integrative, Thematic Field Study

Through the experience of offering forty-two overseas field study programs, I have personally learned from both successes and failures about the opportunities and challenges in implementing successful, integrative, thematic field study programs. In reflecting on field study programs that were less successful, two problems stand out. The first is the failure of faculty members to effectively work together to foster common learning goals. Instead co-instructors often compete for lecture and field excursion time or present redundant or contradictory material due to a lack of communication and coordination. Such a lack is, at best, inefficient and, at worst, confusing and counterproductive for students.

A second problem arises from faculty members who fail to adapt to the experiential approach of teaching in the field, by engaging students in the environments they are studying, and instead rely on traditional classroom-based approaches (particularly lecturing). Past courses have found students toiling in their lodging to complete assignments and papers in the same ways they produce them at home (e.g., relying on secondary sources online). Such an approach fails to take advantage of the field experience and reduces students' direct engagement with the subject matter. Assignments in these instances are often disconnected from one another, focus on narrow subject matter, and miss the opportunity to grapple with the complexity of issues on the ground. Using a traditional classroom lecture approach and disconnected course requirements defeats the purpose of going into the field in the first place, and fails to justify the cost and effort involved in such courses. This is not to say that research papers, classroom lectures, and other traditional teaching methods have no role in field programs. Rather, I have found that these methods are more effective and relevant if they are completed primarily as reflective, post-field program submissions or as pre-course requirements (to which I return in a moment). Further, while some reflective writing and discussions are important to undertake during field study programs, it is important to integrate these processes of reflection into the assignments, field journals, and field excursions rooted in the experience in the field. In situ lecturing can be highly engaging using the field setting as the backdrop (instead of a PowerPoint presentation). From my experience, relying on traditional teaching methods while in the field often results in poorly completed and rushed assignments; exhausted, frustrated students and faculty; and dissatisfaction from students with the program in general.

In the redesign of our field schools, we established a week-long pre-program session. In my earlier involvement as a teaching assistant and later as an instructor on field study programs, I saw that students were lacking in basic background knowledge of our destinations. They often lacked an understanding of fundamental geographical concepts and theory, methods of fieldwork in geography, comprehension of how to approach experiential learning as distinct from learning in the classroom, and a sense of how to integrate various streams of knowledge. Shurmer-Smith and Shurmer-Smith (2002) argue that the "teacherly activities" should happen "well before setting off" (166) and that the leader needs to set up the excursion beforehand "in such a way that the students are empowered" (165).

In the pre-session, we provide students with background readings on the theme of the program and relevant geographical theory, as well as on the historical, political, and physical geography of the regions we will be visiting. Readings are available in the winter term so students have access to them prior to the pre-session week. We provide integrative assignments in the pre-session that have elements of human, physical, and urban geography similar to the assignments they will do during the field program to familiarize them with this integrative approach to examining problems or issues. For example, student groups are often assigned a single city or urban region and tasked with researching important cultural, social, and environmental elements of that location. They are then asked to draw connections among these elements. For a particular city, one student group member might examine how topography influences public transportation or land use. Another might look at climate, particularly seasonality, and how it influences local food production. A third might examine how existing mass transit systems and bicycling accommodate material transport, and a final student might look at municipal zoning principles, analysing how they came to be applied and whether they serve the requirements identified by the first two students. These kinds of integrative assignments help students recognize the interrelated social and physical processes co-producing the situations they then encounter in the field. The pre-session provides students with considerable background to the destinations and the skills required to allow them to hit the ground running in their fieldwork assignments and field journals. This background enhances their overall experience on the field study program and ensures they have the opportunity to make the most of the limited time available. We constantly assess the pre-session work and experiment with new ways of delivering course material.

Beyond preparing students through the pre-session component, the success of the integrative, thematic field study approach involves engaging coursework during the travel segment. A principal task for students is to keep effective field notes capturing observations, critical insights, and reflections. In their field notebooks and journals, students record observations; describe spaces and places; record details of location, inter- actions, conditions, demographics, and so on; try to recognize changes; note observed personal reactions; and reflect on affective aspects. Note- books include written reflections, but also imagery, sketches, photo- graphs, and other visual aids, which are often effective ways of detailing interconnections.[1] The notebooks afford opportunities for future reflec- tion, including tracing changes in one's opinions or feelings. The goal has been for students to begin with general observations and, through trial and error as the program progresses, transition to observations that are more detailed, specific, focused, and defensible. Written reflection can then lead to insightful discussion, where hypotheses can be sub- jected to further scrutiny and questioning. At this point connections and comparisons can be made between places, cities, environments, and stu- dents' home communities.

Along with keeping the field notebook, students are tasked with com- pleting assignments that highlight integration and place-embeddedness, another element that we see as critical to the success of integrative, thematic field study. As an example, on an excursion to study the reconstruction and recovery of communities in Southern Thailand affected by a tsunami, students undertook a complex evaluation of impacts: on tourism infrastructure, local culture, and the economy of the region as well as on the environment and sanitation infrastructure that was rebuilt (or not). They also considered the overall risk to the population from the resulting reconstruction and recovery processes. By encompassing all of these multifaceted variables, the assignment tested the students' ability to pull together and integrate understand- ings from a variety of sources, (sub-)disciplines, and previous courses they had taken.

Beyond coursework, successful team-taught field programs involve commitment and mutual respect, both personally and professionally, for each discipline and faculty member offering the field program. To be able to effectively deliver an integrated program, faculty must engage in an open and ongoing dialogue regarding all aspects of itinerary plan- ning (why are we going there?) and curriculum development and inte- gration (how do the subjects relate to each other?). A clear focus on how to make the field experience most relevant is necessary during the itin- erary and curriculum-planning process and includes reflection on how best to make connections between theory, concepts, and multifaceted approaches to the problems or issues under study. In my most successful field programs (measured in terms of student evaluations and personal reflection), collaboration between faculty was fundamental to achieving integration of subject matter, focus of the itinerary, and in some cases, course requirements.

Encouraging both faculty and students to be reflexive is also important to our efforts at integrative field study programs. Based on his field programs in Singapore and Malaysia, Glass (2014) emphasizes the importance of reflecting on one's social positionality in the field. We encourage students to reflect on the often-hidden assumptions and value-laden language they may use to describe the places they visit. For instance, when we describe places as "sketchy," "dirty," or "crowded," what does this language reveal about us? Do other (local) people seem to have the same reaction to places as we do? We encourage students to reflect as much as possible on how their backgrounds, previous experiences, and identities shape how they see a place. In the oil-affluent city of Calgary, Alberta (where most of our students come from), this can be particularly problematic. Often economically privileged students – who have more of an opportunity to participate in an international field experience – have no first-hand experience with poverty, the density of larger cities, or environmental problems they may directly encounter. These often uncomfortable encounters with the previously unknown are exactly why field schools are valuable. It can be a huge revelation for students to see up close the challenges and problems we face globally and to go through the process of tracing the source of those problems to their own lifestyle or society. Whether it be the impacts of climate change attributed to the economic engine of the oil sands in Alberta or alternatives to urban sprawl and distinct concepts of livability, these reflections can lead to ideas that may improve the quality of life or social aspects of their own community. A more self-aware, reflexive approach to field learning directly helps in the integration of physical and human-related concepts and interactions by addressing not only human cultural considerations, but also aspects of the natural environment and the effects we have on the environments we visit.

Case Studies

Our approach to field study can be understood through specific examples. The following three case studies from programs in Italy, Thailand, and Germany highlight the value of integrative, thematic field study.

Bringing Three Disciplines Together
to Read a Hazardous Landscape in Italy

The University of Calgary's Earth Science Field School to Italy, offered
for the first time in 2016, is an example of how collaboration across disciplines can work. The program involved a unique partnership between
archaeology, earth science, and geography with a substantive focus on
hazards and urban and natural environment processes. From the earliest
stages of the program's development, the other instructors and I worked
together to integrate the delivery of course content and assignments.
The result was a program that afforded a multi-perspectival approach to
understanding a famous Italian landscape. The focus of the trip was
the 79 AD eruption of Mount Vesuvius at Pompeii and Herculaneum.
Through the lens of earth science and geography, students came to
appreciate the physical mechanisms and the magnitude of the eruption that destroyed both these cities. Through the lens of archaeology,
students learned about the ancient urban environment and the social
and political structure of society at the time. By integrating assignments
with components from all three disciplines, students obtained a much
richer understanding of the catastrophe. Visiting the archaeological sites
of Pompeii and Herculaneum and climbing Vesuvius, students could
come to appreciate the magnitude of this ancient disaster and imagine
its effects on culture, society, and the environment. Students could also
gain a clearer understanding of the current threat of such disasters on
this region of southern Italy and begin to contemplate impacts of disasters closer to home, such as the potential for a Cascadian Megathrust
Earthquake on the west coast of North America.

The Italy field study program also underscored the importance of
reflexivity and recognition of one's positionality as a crucial aspect of
understanding complex and dynamic places. Taking into account
advances in feminist and qualitative research in unpacking researchers'
roles in the field, positionality reflects on the knowledge construction
process as being socially and politically formulated by the complex axes
of identity, one's position entering the field, and the power relations that
take place through fieldwork data collection processes (England 1994;
Glass 2014; Katz 1996; Nast 1994; Rose 1997; McKinney, this volume;
Vibert and Sadeghi-Yekta, this volume). These considerations should
be raised explicitly in pre-session meetings, assignments, or courses as
a means of recognizing that both students and faculty going into the
field inevitably bring their biases, preconceptions, and world views to the

Figure 2.1. The Cloaca Maxima, Tiber River, Rome (Aaron Williams 2016).

environments they are observing. While this acknowledgment of positionality and reflexivity is becoming more commonplace in the social
sciences and humanities, such awareness is equally important – and less
often explored – for students studying the natural sciences. For instance,
field excursions focusing on natural landscapes involve visiting places
where people and the environment may be significantly affected by our
presence.

As an illustrative example, in Rome we led a field excursion to the
ancient sewage system of the Cloaca Maxima along the Tiber River that
still drains portions of the city's runoff into the river today. Although the
intended academic focus of the excursion was one of the earliest sewage
systems, its extensive ancient engineering, and current hydrological processes, we inadvertently entered the environment of the homeless population who live along the river (see Figure 2.1). Being mindful of our
positionality and privilege, faculty and students recognized our potential
intrusion into the living space of others and the need to be respectful,
for example, in terms of (not) taking pictures, being aware of noise levels, and otherwise ensuring our presence was accepted. Of course, our
commitment to integrate social and physical science inquiry compelled

us to expand our focus beyond that originally intended to adopt a richer
political ecological perspective on the broader social dynamics that had
rendered this population homeless.[2]

Understanding the Impacts of "More-than-Natural" Hazards in Tsunami-Impacted Thailand

Running a field study program within a region that has experienced a
natural disaster such as the 2004 Indian Ocean Tsunami is a complicated
endeavour. An integrative, thematic approach affords an opportunity
for students to grapple with this complexity. This field school is offered
exclusively through the Geography Department but, once again, it represents conscious integration – in this case of physical and human geographical inquiry. The program challenges students to understand the
complex constellation of social, political, economic, cultural, and environmental mechanisms that have produced the post-tsunami landscape,
taking bearings from theories of political ecology, sustainability, and
environmental management. The goal of the field study is for students to
be able to critically assess the current situation for local communities, the
mixed results of reconstruction efforts, the political context of aid distribution, overall landscape recovery, and the potential for future hazards.
Students come to recognize the "more-than-natural" quality of so-called
natural disasters (Wisner and Walker 2005; Williams and Rankin 2015).

Through an extensive literature review prior to entering the field and
through guidance from faculty in the field, students come with a comprehensive background to the tsunami event in the regions of focus. Using
assignments and field observations that combine human, cultural, urban,
and environmental focuses, students are tasked with making connections
between environmental degradation before the tsunami from resource-
based industries, how these environmental impacts made communities
far more vulnerable to the tsunami, and what the outcome of reconstruction and recovery is in these locations. Wisner and Walker (2005) describe
the tsunami impact and death toll in Thailand as largely a "man-made
disaster," where resource industries of tin mining and rubber tree plan-
tations along with tourism development within the hazard zone made
people more vulnerable to the event (Williams 2013). Students directly
observe how locations where the mangroves and rainforest had been left
intact prior to the tsunami were more resilient, leading to far less loss of
life, while environmental degradation caused by industrial disturbance
made people and communities particularly vulnerable. Students meet

people in communities who survived the event and hear first-hand about their experiences of loss, the aftermath, reconstruction, and their lives today. Having first studied the literature illustrating how cultural and societal structures both pre- and post-tsunami influenced reconstruction, aid distribution, and recovery of communities, students can observe how these factors are represented in the rebuilt environment and how the rebuilt infrastructure has left people vulnerable to a future tsunami or storm surge event. Through this integrative, thematic approach to field studies, students see beyond the surface of the reconstructed landscape, understanding the social, cultural, economic, political, and environmental influences (the political ecology) that led to the current texture of everyday life and the built environment of the post-tsunami landscape.

Observing the Politics of Sustainability on the Ground in Freiburg, Germany

In multiple field programs we have included Freiburg, Germany, a city widely renowned for its commitment to sustainability and livability initiatives. Our visits have centred on the neighbourhoods of Vauban and Rieselfeld. Over the fifteen-plus years since they have been constructed, these communities have drawn in many researchers interested in studying their model for sustainable planning, livability, renewable energy, energy efficiency, community interaction, and transportation connectivity (Newman and Beatley 2009; Schroepfer and Hee 2006).

In our first visit to the communities in 2008, we took the sustainability of these communities to be self-evident, and we accepted fairly uncritically that these were best-practices examples on the cutting edge of urban design; however, as we made more on-the-ground connections with locals and scholars at Freiburg University, we developed a more complex, multi-perspectival view. From an environmental perspective, the communities seemed ideal: they boast low ecological footprints based on district heating, power generation, plus-energy housing, minimal car use, and density. However, the sustainability of these communities from a social and economic perspective seemed much more questionable, especially in the case of Vauban. Taking the perspective of political ecology, the issues that began to emerge for us in Freiburg were not as simple as highlighting sustainability; there were also many political and economic issues that came to the forefront. In multiple subsequent visits, we discovered that Vauban was departing quite dramatically from the original intentions of planners and the way it is depicted in many published scholarly accounts. Originally intended to be a middle-class community for families, Vauban has become increasingly expensive, unaffordable for younger families and non-affluent individuals. This change in demographics has led to a more exclusionary environment that students could recognize in the ubiquity of signs designating spaces as "private," demanding people to "keep out," and prohibiting photography, even in supposedly "public" places. Interestingly, some of this signage is a result of the numbers of people visiting this community, a kind of theme park for sustainability.

The contradictions of the place became starkly evident to us in one of our visits when we learned about the eviction of the residents of a caravan (mobile home) park located within the community. This park was once central to the fabric and identity of the community and lived in harmony with the residents of Vauban. The caravan park pre-dated the building of Vauban (Schroepfer and Hee 2006). In the planning and construction of Vauban, it was agreed among City of Freiburg officials and the original residents of Vauban that the caravan park residents could stay. With the growing affluence of the community, along with the emergence of Freiburg as a "green capital" destination (as marketed by the city government), it was decided that the caravan park residents would be evicted to make room for a "green" hotel to house visitors to the sustainable community (Mössner 2015). Evidence on the landscape of this upheaval was apparent throughout the community. Signs and graffiti contested the actual sustainability of the community. For instance, Figure 2.2 depicts a sign at the entrance to Vauban in 2009, during the run-up to the eventual forcible eviction of the caravan residents and protesting plans to remove the caravan park. It reads "green capitalism is a lie." Our students independently observed these signs and reflected on their meaning in their assignments and fieldwork. Through this experience, they began to question the extent to which Vauban really met the criteria for a sustainable community. Our students debated whether Vauban and other sustainable communities were truly designed to be socially sustainable or whether they were simply eco-settlements for those elite in society who could afford to live there, and we discussed the resulting implications of such questions. Finally, evidence of similar issues and conflicts in our own city became impossible to ignore and a broader perspective on the theme of "sustainability and gentrification" was the outcome. What had originally been an exercise in examining and analysing places "out there" had inspired new questions and perspectives through which to see our own home communities.

Figure 2.2. "Green capitalism is a lie" (Aaron Williams 2009).

Part of what makes this kind of field exercise valuable is the personal connections developed over the years between Canadian and German faculty members. On the surface, it would be difficult to see the change in demographics, public policy, and culture in communities like Vauban. However, interactions with colleagues at Freiburg University informed my colleagues and me of the changes that were occurring, and we were able to observe these changes across our yearly visits. The connections we have made over seven field programs to Freiburg have not only provided us with a rigorous and meaningful field study destination for our program, but also helped us establish permanent research connections between faculty that have included multiple publications, conference presentations, and a faculty member from our department completing a six-month sabbatical at Freiburg University. Two aspects of these collaborations stand out as particularly unique and meaningful. Through his involvement with our field study program, a faculty member in the Department of Geography at the University of Calgary has become one of the leading experts on critical sustainability research in Freiburg. Second, students are able to observe the positive aspects of sustainability within these communities and yet gain a more critical perspective of sustainability in Freiburg through learning alongside faculty who research these specific communities.

Concluding Reflections

The intention of this chapter was to highlight the value of an integrative, thematic approach to field study. My own experience leading field schools has shown me that such multidisciplinary field schools – whether co-taught by faculty from separate departments (such as in in the University of Calgary's Earth Science Field School to Italy) or taught within one specific department in an interdisciplinary way (such as the University of Calgary's field school to Thailand taught in the Department of Geography) – provide students with unique learning opportunities that are difficult to reproduce in the classroom. We can inspire students to celebrate and integrate different ways of apprehending the world and the complex processes shaping and transforming the interconnected human and natural environments we encounter out there.

NOTES

1 Through advances in digital photography and photo editing software, we are now much better able to integrate photography into the traditional field notes methods. For example, students complete photo essay assignments making connections between the natural and human environment or create digital story maps. Photos and sketches can also be used to inspire discussion. As an example, a student sketching tree leaves in public parks can spark discussions of landscaping philosophies, biodiversity, green space allocation, and the "urban wild."

2 Likewise, when in human-focused field studies, considerations of the natural environment can be emphasized with positive results. An example of this is the use of resources that field study programs may use: the amount of waste we leave behind through our presence and the water and energy resources we use. This becomes particularly clear and important in visits to developing countries we visit with limited resources and infrastructure. Students can then consider their place within the environments they enter as being natural environments even if they are altered by human presence. Exploration of this topic leads naturally into yet another aspect of integrative field studies, which is considering the place of humans within the environment as a whole.

REFERENCES

Cantor, Jeffrey A. 1997. *Experiential Learning in Higher Education: Linking Classroom and Community.* Washington, DC: Association for the Study of Higher Education-ERIC.

England, Kim. 1994. "Getting Personal: Reflexivity, Positionality, and Feminist Research." *Professional Geographer* 46 (1): 80–9. https://doi.org/10.1111/j.0033-0124.1994.00080.x.

Friedman, Thomas L., and Joanne J. Myers. 2005. *The World Is Flat: A Brief History of the Twenty-First Century.* Updated and expanded edition. New York: Macmillan.

Glass, Michael R. 2014. "Encouraging Reflexivity in Urban Geography Fieldwork: Study Abroad Experiences in Singapore and Malaysia." *Journal of Geography in Higher Education* 38 (1): 69–85. https://doi.org/10.1080/03098265.2013.836625.

Hains, Bryan J., and Brittany Smith. 2012. "Student-Centered Course Design: Empowering Students to Become Self-Directed Learners." *Journal of Experiential Education* 35 (2): 357–74. https://doi.org/10.5193/JEE35.2.3576.

Katz, Cindi. 1996. "Expeditions of Conjurers: Ethnography, Power, and Pretense." In *Feminist Dilemmas in Fieldwork*, ed. Diane L. Wolf, 170–84. Boulder, CO: Westview.

Mark, Jason. 2012. "Natural Law: From Rural Pennsylvania to South America, a Global Alliance Is Promoting the Idea That Ecosystems Have Rights." *Earth Island Journal* 27 (1): 40–6.

McDonald, Garry W., and Murray G. Patterson. 2007. "Bridging the Divide in Urban Sustainability: From Human Exemptionalism to the New Ecological Paradigm." *Urban Ecosystems* 10 (2): 169–92. https://doi.org/10.1007/s11252-006-0017-0.

Mössner, Samuel. 2015. "Urban Development in Freiburg, Germany – Sustainable and Neoliberal?" *Die Erde* 146 (2–3): 189–93.

Nast, Heidi J. 1994. "Women in the Field: Critical Feminist Methodologies and Theoretical Perspectives." *Professional Geographer* 46 (1): 54–66. https://doi.org/10.1111/j.0033-0124.1994.00054.x.

Newman, Peter, and Timothy Beatley. 2009. *Resilient Cities: Responding to Peak Oil and Climate Change.* Washington, DC: Island Press.

Robbins, Paul. 2012. *Political Ecology: A Critical Introduction.* 2nd ed. Chichester, West Sussex: Wiley-Blackwell.

Rose, Gillian. 1997. "Situated Knowledges: Positionality, Reflexivites and Other Tactics." *Progress in Human Geography* 21 (3): 305–20. https://doi.org/10.1191/030913297673302122.

Schroepfer, Thomas, and Limin Hee. 2006. "Sustainable Urban Housing: Design Ideals and Ideas for Vauban." *International Journal for Housing Science and Its Applications* 30 (4): 281–92.

Shurmer-Smith, Louis, and Pamela Shurmer-Smith. 2002. "Field Observation: Looking at Paris." In *Doing Cultural Geography*, ed. Pamela Shurmer-Smith, 165–75. London: Sage.

Universities Canada. Universités Canada. 2014. "Internationalization at Canadian Universities: Quick Facts." Accessed 21 Feb. 2017. http://www .univcan.ca/wpcontent/uploads/2015/07/quick-facts-internationalization -survey-2014.pdf.

University of Calgary. 2013. "Becoming an International Hub: Highlights from the University of Calgary International Strategy." Accessed 21 Feb. 2017. http://ucalgary.ca/research/files/research/becoming-a-global-intellectual -hub.pdf.

University of Calgary. 2016. "Summary of Undergraduate Enrolment." Accessed on 30 April 2017. https://oia.ucalgary.ca/files/oia/2015-16fb-2-3-sum-x-ug .pdf.

Vaira, Massimiliano. 2004. "Globalization and Higher Education Organizational Change." *Higher Education* 48 (4): 483–510. https://doi.org/10.1023/ B:HIGH.0000046711.31908.e5.

Williams, Aaron. 2013. "Assessing the Sustainability of Tsunami-Impacted Communities of Thailand's Andaman Coast: An Institutional Ethnography." PhD diss., University of Calgary, Calgary, Alberta.

Williams, Aaron, and Janet Rankin. 2015. "Interrogating the Ruling Relations of Thailand's Post-Tsunami Reconstruction: Empirically Tracking Social Relations in the Absence of Conventional Texts." *Journal of Sociology and Social Welfare* 42 (2): 79–103.

Wisner, Ben, and Peter Walker. 2005. "Getting Tsunami Recovery and Early Warning Right." *Open House International.* 14 April. Accessed 20 Feb. 2017. http://www.openhouse-int.com/abdisplay.php?xvolno=31_1_6.

BEING PART OF SOMETHING BIGGER ...

One morning while we were all getting prepared for a busy day, our attention was brought to the most painful heaving cries ringing out into the village. A number of members of our class came out to our balcony, which overlooked the village. We could see Jaiyama, an elder woman, who was a main actor in the company, grievously falling into the arms of other elders. We didn't know what was happening, but we sensed there must have been a death. We had witnessed a few deaths in the village during our stay up to this point. Later that day, Gus informed us that Jaiyama's only surviving son had passed due to his struggle with alcoholism – a struggle that was directly addressed in a key story Jaiyama had shared for the play the company was to perform in one week. Jaiyama was given the option of not performing the story, but in the end she decided to go ahead and perform. Not only was this a rare opportunity to see a performance of a true and personal story by an actor, but the fact that she had the courage to openly share her story with her community while being in the thick of grieving her son's life was life-changing to witness. Being witness to Jaiyama's strength and openness made me realize I was part of something much bigger than I had anticipated.

Kathleen O'Reilly
Participant, University of Victoria's
Applied Theatre Field School in India

3

The Enlivened Classroom: Bringing the Field Back to Campus

NAKANYIKE B. MUSISI

Introduction

We are at an exciting juncture in the provision of higher education in which core values and practices are all up for interrogation. This is, in part, because field schools, once the exclusive preserve of a few select disciplines such as anthropology and geography, are now increasingly offered by a far wider range of disciplines. This development has generated a number of insightful reflections on various aspects of field schools and the possibilities they offer to enhance on-campus pedagogy; see, e.g., Mycroft (2009), Kolb and Kolb (2009), Smith (2011), Slavich and Zimbardo (2012), Rennick and Desjardins (2013), Che, Spearman, and Manizade (2009), and Mitussis and Sheehan (2013). Coming from the discipline of history, I add to those reflections by drawing on my participation in two unrelated programs administered by the University of Toronto – the Summer Abroad program, administered through Woodsworth College, and the International Course Module program, run through the Faculty of Arts and Science – and consider how these programs have transformed my on-campus teaching.

The chapter is organized into three sections. The first section provides background information, describing the field set-up and activities of each program. The second discusses some of the strategies I have adopted in my on-campus pedagogy that draw on the experiences of the two field schools. In the final section I reflect on the lessons learned, including both ecstasy and adventure,[1] and some of the challenges the students and I encountered in the process. I argue that transferring the lessons learned from the field school experience into on-campus pedagogy disrupts the commonplace at many levels. It permits and enhances

more active student engagement in learning, maintains an epistemological curiosity in students, and unsettles the problematic student-teacher
hierarchy and other binaries, such as developed West/underdeveloped
Africa, us/them, oppressed/oppressor (see the chapters by McKinney
and by Vibert and Sadeghi-Yekta in this volume) while, at the same time,
stimulating multilayered reflection about and action in the social world –
with a goal of contributing to its transformation. I contend that building
students' confidence facilitates their ability to make connections to the
larger social worlds that they inhabit in ways that go beyond what they
would typically gain from a traditional lecture-based pedagogy.

Field Set-up and Activities of the Summer Abroad and International Course Module Programs

The Summer Abroad program is designed with the goal of enriching
students' academic lives by providing an exciting and educational international experience of three to six weeks. The International Course
Module program, by contrast, incorporates an intensive international
experience into the framework of existing undergraduate courses.
Travel is scheduled to coincide with Reading Week (mid-February). I
have to date participated in three rounds of the Summer Abroad program and one International Course Module over a period of four years.
These excursions have in total involved close to fifty students.

The course I offer in the Summer Abroad program is called "Conflict
and Community in Africa," and it takes students to Kenya for three weeks.[2]
This short-term off-campus study program is designed as an intensive
inquiry into the causes, consequences, and especially possible responses
to conflict in Africa. The overall objective is to introduce students to the
complexities of conflict, peace, and development work in Africa while
at the same time developing in them an appetite to engage in the world
around them. The course is organized into two parts, both of which
unfold on the ground in Kenya. The first part investigates the causes of
violent conflict in Africa and consists of lectures and seminars on competing conceptions of conflict, such as primordialism (ancient hatreds),
constructivism (social and political construction and exploitation of ethnic identity), and environmental scarcity or resource curse. Visits with
Nairobi-based resource people help us reflect on how well these academic explanations account for violence in Kenya and elsewhere. The
second part explores possibilities for moving beyond violence. Through
lectures and seminars, students consider contentious issues such as the

role of the international community in humanitarian aid and intervention and the challenges of peacekeeping and peacebuilding. Students
also get opportunities to engage with non-governmental organizations,
local researchers, peacekeepers, diplomatic officers, academics, and
ordinary Kenyans through site visits, including community walks or working on specific tasks, lectures, and seminars. Throughout, students work
in groups to apply the theoretical frameworks and concepts to a variety of
contemporary and, in most cases, ongoing conflicts in Africa. The course
also has a structured service learning component that addresses human
and community needs in the Maasai Mara. This component is organized
and led by our host organization, "Me to We – Free the Children."[3]

Conversely, the International Course Module is part of a fourth-year
seminar course that uses a biographical approach to investigate the lives
and times of elite African women in the twentieth century. We examine oral history and life history as historical methods while critiquing
the knowledge and politics they both evoke. Discussions are organized
around the auto/biographies of three women and one couple. One
woman is from West Africa, two from East Africa, and the couple is from
South Africa. Two of these women and the couple have passed away;
only Miria Matembe of Uganda is still alive. The International Course
Module is organized around her biography and her historical moment
in Uganda's history.

Although the programs differ in structure and aims, they share similarities. Both attract a heterogeneous body of students. For instance, my last
Study Abroad program (2014) resembled the United Nations, with twelve
students drawn from eleven countries. Likewise, students were drawn
from different disciplines and social classes, yet they tended to generally have similar motivations for being involved in the programs, such as
gaining an international experience, attaining a different perspective of
the world, gaining new and/or enhancing their life skills, and learning
in a new and different environment. Some structural elements were also
shared. For instance, in both courses we convened every few evenings to
share thoughts and reflections on the previous days' experiences. These
meetings, whether in Kenya or Uganda, often included strong reactions
and emotional outbursts. I recall one evening over a campfire when one
student confronted another for "standing in [their] way to learn." The
outburst was caused by the accused student uncompromisingly challenging one of our guest speakers, claiming that she was a "white Maasai."
The same student had agitatedly protested what she saw as a "saviour
mentality" in the "Me to We – Free the Children" display and narrative

around the old school and new school buildings in the Maasai Mara. She had also objected to what she saw as exploitation of the "Mamas"[4] when, through our host organization, we "invaded" their homes. Our activity of fetching water for host families she saw as "pretentious and lack[ing] depth." There was no one single blueprint to help us negotiate such differences. Occasionally, it was enough to remind students that while we are in the field, what we see, hear, taste, feel, and experience may be uncomfortable and disorienting, but we must not lose sight of the reason we are there. I would emphasize the importance of building a safe place, which for us meant that we were all good listeners and respectful of all opinions, even those with which we disagreed, while at the same time adopting a critical view in order to learn about the situation in question and about ourselves. It was important not to "domesticate difference" (Burbules 2000, 261) but rather to regard such tensions as part of the vital building blocks for our intellectual collaborative intimacy (Mitussis and Sheehan 2013). In spite of such differences of opinion and politics, strong bonds emerged among members of the group – sometimes with hugs and reassurances that all would be fine.[5] Above all, on-the-ground coordinators, my teaching assistant, and I remained attentive in monitoring students' emotional and physical state. That vigilance was mutual; often students looked out for me and made sure that I did not skip a meal because of the numerous programmatic demands I faced.

Procedurally, by virtue of being an integrated on- and off-campus course, by the time we left for Uganda, students in the International Course Module had formed a learning community and gained substantive (academic) understanding of the host country. Upon our return, students had the opportunity to think about whether and how their views about Matembe changed or were reaffirmed. Students also had the chance to consider the contradictions they saw, such as the view of Matembe as the adept feminist working for social justice for women but, at the same time, the view of Matembe as a very "polite" homophobe. On the whole, Miria Matembe and the others they interacted with had a large impact on the students. For instance, my teaching assistant who at the time was pursuing a PhD in medieval history quit her program, confiding in me that "if women who had no PhDs could have such tremendous effect on society," it made no sense to her to continue to study medieval European history. Her immediate supervisor was disappointed and attributed her decision to quit to the fact "she must have gotten 'sick' in Uganda."[6] Rather, the student seems to have exercised her right to act on her cognitive dissonance (Difruscio and Rennick 2013).

Both in Kenya and Uganda, students' positions and responsibilities shifted regularly. Their roles ranged from taking on group leadership, to acting as motivational boosters, to role-playing as "devil's advocate," through tasks such as listeners, note takers, presenters, and discussants of the day's key issues appearing in the national daily newspapers and relating to our course of study. Neither was my role stable. It oscillated between instructor, fellow learner, and team leader and at times extended to mother, sister, and friend. I was able to draw on my vital connections in Uganda and Kenya for our intellectual quest, and I did all the communications for appointments and other arrangements in advance. In that sense, I was the team leader; nonetheless, my role from that point forward took a different turn. We stayed together in the same dwelling, and my door was just next to theirs; we ate and did everything else together. I was just one among the group. And although at times I would assume authority or steer students towards understanding a particular point practically, a casual observer would have been hard pressed to identify me as the professor. By putting myself in this liminal position, I released the hierarchical rigidity often experienced in on-campus teaching. Students undertook some of my roles as I took on parts of theirs[7] (Wegerif 2007, 4). Situating myself as a fellow learner, rather than the expert, changed my role as a teacher. I became a counterpart in the process of learning (McCaleb 1998; Wink 2005). This affable and open relationship with the students in the field allowed them to get to know me better, to crack jokes with me, and sometimes to ask me very personal questions. Suffice it to say that at best our roles, daily lives, and interactions were marked by ambiguity, which was at times exhilarating and adventurous and at times dissonance triggering.[8]

Bringing It Back to the Classroom

This rich experience in the field has significantly influenced my on-campus teaching. Rearticulating on-campus learning outcomes has meant putting emphasis on a number of analytical skills beyond the usual critical thinking skills and paying attention to the formation of values and habits of mind. Combined, these skills include knowledge application, problem solving, decision-making, creative thinking, information literacy, independence, curiosity, and learning about one's self. Above all, this has meant seeing students' and my role in the classroom differently. With this in mind, the ultimate goal has been to increase students' inquisitiveness in the acquisition of knowledge while at the same time accomplishing two other things. First, a new syllabus goal aims at building on and interrogating students' deeply held assumptions and lived values, and discovering, cultivating, and celebrating their untested potential strengths, skills, and endowments. Second, after years of concentrating on content in my history classes, my emphasis has shifted to the processes of learning and teaching. This has meant changes in the following four key areas: (1) a shift towards classroom techniques that create a community of learners who are directly engaged in their learning; (2) changes in the course delivery model, particularly how we work with materials, texts, guest speakers, and films to boost our investigational capacities; (3) changes in the "hows" and "whys" of assessment and evaluation; and lastly, (4) a decentring of authority by altering my position in the classroom. (For flipping the classroom, see Abeysekera and Dawson (2015), Barkley (2009), Mok (2014), and Strayer (2012).) While alternative pedagogies in higher education, such as critical pedagogy, active learning, and student-centred learning are by no means unique to off-campus study programs, it was my own experience leading short-term study abroad programs that helped me realize the potential and value of these pedagogies, which then led me to intentionally integrate them into my on-campus history courses as well.

Creating a Community of Learners

Transformative success of the field is partly attributed to the context/setting within which iterative engagement and learning takes place (Rennick and Desjardins 2013; Mitussis and Sheehan 2013). There was no question that the groups' social dynamics in Kenya and Uganda were fundamentally shaped by the specific situational settings. The question for me became the following: how do we replicate this environment, in an on-campus setting, to nurture curiosity and openness to learning that could lead to a transformative experience? Basically, how do I increase students' inquisitiveness and development for transformative learning (Brewer and Cunningham 2009)? The answer was to let the students discover and take control of their learning by allowing them to be challenged beyond their usual expectations, to express their emotions, to work more collaboratively, and to celebrate their small discoveries and victories (Wink 2005; Buckingham 2003; Che, Spearman, and Manizade 2009; Freire 1970). This approach would mean becoming particularly sensitive to and investing in classroom dynamics.

Although classes are scheduled for two hours and we meet only once a week, I have included a number of activities to increase points of contact and interactive engagement – in hopes of recreating a measure of the intimacy and critical engagement present in the field. Group work has become a standard practice in my classes, and physical classroom atten- dance is mandatory. Usually, we end the class by coming back into ple- nary to share each group's main points and see what consensus emerged, and what differences arose, around the reading and the topic of the day. In the out-of-class group activities, I ask students to keep a log detailing the time put into such exercises and to note their accomplishment and challenges and how they went about solving them.[9] When groups were to lead discussion, they worked hard to engage the class.

In the 400-level course, the two-hour session is divided into three parts. The first part involves reviewing the material. It is both descriptive and performative and directed at understanding the content characters and setting using different games such as *Jeopardy*, flip quiz, randomized bingo cards, vocabulary space race, and so on. Scores are kept, and the win- ning team is rewarded. In the second part, with clearly worded questions, the group leading the discussion has the class reflect on the readings. This section often provokes different reactions ranging from anger to affirmation, rejection, or even harsh criticism. For example, one student shared a similarity with Wambui Otieno in how her mother was treated by extended family when her father died; another pointed to his identity crisis arising out of his family's moving from one continent to another. In the final section of the class session, we critically analyse how the story was put together identifying the plot and author's intent, biases, strengths, and weaknesses in stringing different elements of the story together, and what is said between the lines. The efforts of the group leading the discus- sion are acknowledged and celebrated with a round of applause.

In the 300-level courses, students are also expected to use the online discussion section of their course management system to continue their reflections within 24 hours after class. Discussions on the course manage- ment system are open, and students are encouraged to read and reflect on each other's entries. Besides these measures, a community has been created by ensuring that students co-own the classroom and, above all, that this is a safe space that allows open and curious questions, personal reflections, and freedom to talk or process one's dilemmas or share one's own stories, taking intellectual risks. As in the field, we work to create an environment where playfulness could emerge without fear and could be cognitively challenged but not judged and maligned; see Che,

Spearman, and Manizade (2009, 103), Buckingham (2003, 58), Bruner (1996), Gokhale (1995), McKeachie et al. (1990). The content review games (as described above) often do the trick. Akin to what is described by Joan Wink (2005, 131), at times my enlivened classes have seemed disorderly and possibly too loud for any serious learning to be taking place. Without much persuasion, students have drawn from their personal lived experiences, what they have learned in other classes, or what they have heard or read on social media, as we challenged the accepted narratives on various issues (such as HIV/AIDS or female genital cutting) in Africa.

To ensure that any particular history course I teach should potentially lead to students' engagement with the world around them, I ask a relatively simple set of questions at the very beginning: Why do you want to take this course? Why is African women's history important to you? The usual answers are couched around expressions such as: "I want to learn more about ..., understand ... know my heritage," and so forth. The next question, which usually takes them by surprise, is: "So what?" Their answers tell me that they want facts, and I applaud them for this inquisitiveness, but I am quick to disappoint them by pointing out that they would probably not get the answers they are initially looking for and that, at times, I will have no answer to their questions – and that besides, there will generally be no one correct answer to the problems we confront. I tell them that the course is as much about them as it is about Africa. It is about them learning and relearning methods of acquisition of knowledge, challenging assumptions and positions, and attempting to explain why a particular problem was presented as it was. I make them aware that the course is a privileged opportunity and space for them to embrace uncertainty and shifting understandings of African history and the world we inhabit.

Changes in the Course Delivery Model

To enrich the students' appreciation of African women's history, I continue to use films and other forms of media, guest speakers, and other materials I have used in the past. But I now use them for a different purpose: not to transmit facts and transparent information, but rather for discursive purposes. I ask students not to take anything they see or hear at face value but to ask deeper questions, for example, to look deeper at images, assumptions, word choices, overt and hidden agendas, and to attempt to understand the short- and long-term implications of what is presented. I am concerned about content but also about how that content has been packaged and delivered. This approach has resulted in sharper observational and analytical skills and, potentially, in higher levels of comprehension. In the non-seminar courses, classes are organized around a key topic, and each begins with me giving an introductory presentation focused on the issue at hand. After this, the class breaks up into smaller groups for discussion. The intensity of discussion fluctuates from group to group, depending on the topic, how students work with the week's texts, and their prior knowledge. The dialogic approach that forms the core of our learning is based on my and the students' active, collaborative discussions (Bruner 1996; Brookfield and Preskill 1999; Lipman 2003; Alexander 2006a, 2006b). Within this dialogic space, learning is potentially transformative because it is experienced as a process not of knowledge transmission or construction but of knowledge exploration and questioning – where what is presented is not the ultimate truth or final word but one answer among several possibilities: what was a previously held assumption could be seriously challenged (see McKinney, this volume). For my part, dialogue with the students is maintained through verbal and written feedback and discussion in and outside of the classroom. For instance, to students who might ask me "Where can we find information on nudity as a form of African women's resistance to colonial rule?" my response would be "Where do you think your number one place to search should be?" If a question is asked in class, my first response is "Can anyone answer that question?" Usually, there is one student who has an answer, and their answer frequently snowballs into an animated discussion with often a range of more brilliant answers than I would have given.

In the 400-level seminar, a dialogic method that embraces a playful and celebratory format for each class has deepened students' learning more than I could have ever anticipated. Students make meaning of ideas instead of expecting to be given answers or leads. This year, the massive volume of readings did not even seem to be a cause for complaint as usual.[10] The seminar took the format of a game (*Jeopardy* or whatever form the students chose) to plough through content, followed by a discussion of questions, and ending with a celebration. Students rewarded each other with goodies as well as a mark. Each group (consisting of a maximum of four people) led the class for two consecutive weeks. After the first, they received blind peer assessment in addition to my verbal feedback to consider any improvements for the following week. At one point, I thought that after two classes of using the same format, we should make some changes. I was overruled. The class continued with high energy and likely would have appeared to the passerby as electrified anarchy – as exhibited by boisterous laughter and spirited and loud overlapping voices, as well as the movement of bodies (see Wegerif 2007; 2008).

Changes in the "Hows" and "Whys" of Assessment and Evaluation

Assignments have also taken on a new twist as we work together to bring the benefits of the field back to campus. In my larger classes, rather than continuing with the History Department's conventional two written assignments and a test/exam at the end, I have started to assign short formative assignments, often snowballing towards a longer essay of higher quality.[11] And because the learning transmission model encourages students to work cooperatively, some assignments require group work and a group mark. Anxious enquiries (either by email or through office hours) into what is expected for both individual and group submissions are a common occurrence. Cognitive disequilibrium – "confusion, uncertainty and ambiguity" – is often at the heart of this search for clarity (Tennant 2006, 130; Ellsworth 1992; Che, Spearman, and Manizade 2009).

Akin to quick feedback in the field school setting, students get back their papers punctually and with substantial feedback. But in spite of this prompt response, a desire for encouragement, reaffirmation, or tips for improvement cause even "A" students to frequent my office hours to inquire as to how they are performing in the class.[12] I have also adopted other alternative methods of assessment and evaluation. For example, as already noted, group presentations involve peer review and assessment. Working with an assessment tool, the group presenting is given a mark and comments by the rest of the class, and the group itself evaluates how the individual members of the class participated in the ensuing discussion. Students in my seminar class this year made recommendations on how best to apportion marks and which assignments might be combined. Students in the Kenyan and Ugandan Study Abroad and International Course Module programs were given an opportunity for self-assessment. In the spirit of alternative methods of assessment, I also give bonus points to students who meet deadlines or take on additional tasks (for the benefit of the class), or are always in the lead with their reflections on the course management system. Last, but not least, I have challenged my students to look at their assignments and grades as a process not a product, and because my classes are relatively small with a workload of two under-graduate courses a term, I often offer assignment rewrites.

Decentring of Authority

To encourage field school-type productive closeness in an on-campus set-ting, I operate an open-door policy and hold my office hour immediately after class. In such settings, I continue our discussion, asking students questions as well as listening to their feelings and the meanings they bring to a particular problem.[13] I also now make a point of entering the class dramatically: I might make a joke about the weather, smile, high-five a student in front, allow a moment of being "silly," and then bring the class to order. In the seminar class, I sit among the students, rotating my position to ensure that I sit next to each one of them by the end of the term. And although at times I find myself playing an active role in challenging the students' assumptions or previously held views, overall I have significantly reduced my talking time. I am increasingly becoming a fellow learner, a listener or interpreter, who from time to time provoca-tively challenges and opens up alternative readings of experience or of the texts under analysis (Lipman 2003, 87; Wink 2005). I encourage and remind my students to see themselves as learners of today, in an age of unprecedented mobility of ideas and people, where the concentration of knowledge in one person (the teacher) is a relic of yesteryear. They are members of a broad learning community – rather than of the top-down models of the past (Wink 2005).

Challenges and Prospects

Bringing a field school learning atmosphere into the classroom has a handful of challenges. First, there has been resistance by some students to my disruption of the "normative" teaching method, where I stand in front of the class and lecture (pontificate), they sit (quietly), take notes, and perhaps ask a few questions to which I give answers. This is an abdi-cation on the part of some students to take full responsibility for their learning (Biggs and Tang 2007). These same students would often prefer a final examination where they would repeat what has been transmit-ted in class or through the readings rather than incremental, short but intensely engaging assignments. For the first few classes, in addition to underscoring my lack of belief in exams, I reiterate the importance of their active engagement with the course materials (ideas and facts) and interaction with each other and our guest speakers or other resource persons in class.

A second challenge has been cultivating intellectual collaborative intimacy among students (Mycroft 2009; Mitussis and Sheehan 2013). Part of the success of the Summer Abroad and International Course Module programs is that, given the setting, students quickly develop intimacy for each other – in contrast to a setting where they meet once a week for two hours. Granted, ordinarily, intimate connections are in any case not easy to make, but with changes in the pedagogical and classroom techniques (as described above), an environment is built in which "intellectual collaborative intimacy" potentially starts to develop organically. When this occurs, as I witness in my classes this year, the sky becomes the limit for truly dialogic learning to take place. Students enjoy coming to class, and the two hours become a celebration of their discoveries together.

A third challenge is how to circumnavigate the shortcomings of a dialogic pedagogical model (Burbules 2000). For instance, it is necessary to ensure respect for all points of view and deal with those who opt not to participate in this dialogic communication and might feel dispossessed or sidelined. In particular, the challenge is how not to "domesticate difference and resistance" but to see these as a potential starting point for dialogue (Burbules 2000, 261). Moreover, dialogue is not simply a matter of who and what is present in the classroom. Its smooth execution might be constrained by, for instance, interactions before and after class, knowledge from other classes, upbringing, cultural background, and so forth (Ellsworth 1992; Burbules 2000, 269). Rather than seeing these challenges as inhibiting, I am learning to take them as potentially crucial ingredients in making productive cognitive dissonance possible outside of field school settings. Nevertheless, it is still a challenge to effectively deal with students' "disorienting dilemmas" (Brewer and Cunningham 2009, 9; Mezirow 1997, 7; Kiely 2005; Glass, this volume). The attendant challenge has been how to deal with at times powerful emotions or confusion caused by unaccustomed new ways of learning.

The biggest challenge for both my students and myself has been how to balance the demands of total immersion with other commitments. Furthermore, on-campus students and I are rarely in close proximity to each other outside class time. The challenge has been how to make ourselves available to each other without too much collateral damage.

On the positive side, in spite of the challenges noted above, I feel encouraged by the changes implemented and student response to these. I am happier in the classroom than I have been at any time in my teaching

career. The passionate feeling is mutual, as is evidenced by some students' verbal communications and their end-of-course evaluations. Yes,
predicaments have sometimes been profound and resolutions rare. For
instance, one of my male students frequented my office after class and at
other times and told me:

> I am finding it hard to reconcile my belief system with what I am gaining in
> this class – I have never questioned this area of my being.

A female student disclosed to me:

> Professor, what I am learning in this class is very dangerous for my well-
> being – but in a positive manner! I am questioning everything.

Confidence and safety were also exhibited when another student, whose
ideas were consistently challenged by outbursts from his classmates, con-
fided in me, saying:

> I am going to continue bringing the knowledge and ideas I have gained in
> my other classes to test them out in this class ... I realize that whenever I
> utter what I consider to be common knowledge, I get it wrong in this class!

Has "constructive disequilibrium" taken place and have students like
these been empowered (Shor 1992, 129; Che, Spearman, and Manizade
2009; Wink 2005)?

These examples and many more underscore the importance of a safe
and respectful intellectual space as well as nourishing classroom dynamics that permit highs and lows to be expressed without debilitating cen-
sorship. Equally important, these examples and many other expressions
in class and beyond demonstrate that students in these classes co-invested
in the dialogic pedagogical model, where "self and other mutually constructed and reconstructed each other" (Wegerif 2008, 353) and where
"discussion confirmed students as co-creators of knowledge" (Brookfield
and Preskill 1999, 25; Shor 1996). Students found these classes to be
a space where they could test their prior knowledge and assumptions,
reimagine their future engagement with Africa, and cite a combination
of what they have gained in class discussions or feedback from their colleagues as well as myself. All of this combines to make me sanguine that
methods intended to bring "the field back to campus" have a potential of
being positively productive and transformative. Nonetheless, for reasons

similar to those enumerated by Mezirow (1997), I shy away from making claims that all students experienced a transformative experience.[14]

With regard to myself, I had previously taken the quality of being an affectionate and passionate teacher for granted and ignored it as vital pedagogical capital (Mitussis and Sheehan 2013). In the field, it was a matchless asset to encircle my students with confidence and resilience. Reconciling this part of me in my on-campus pedagogy liberates my teaching from the "coldness" of transmission pedagogy. Student evaluations have pointed to my passion for history, affection for students, and the value of my past international experience as crucial in fostering their critical engagement with Africa. For my part, pedagogical experiences drawn from the Summer Abroad and International Course Module programs have been highly enriching, leading me to a more meaningful, imaginative, collaborative, and democratic relationship with my students (Fielding 2001; Tennant 2006; Alexander 2006a, 2006b).

Conclusion

In 2011, before accepting the invitation from the director of the Summer Abroad program to take my first group to Kenya, I struggled with a number of ethical issues. How could I participate in the program without being seen as colluding with the tenets of "exotification" or engaging in a neocolonial project of taking students to Africa to visit "natives" – a kind of "edutourism" that buttresses stereotypes about Africans? (See Kiely (2004), Epprecht (2004), Sichel (2006), and in this volume, the chapter by Vibert and Sadeghi-Yekta.) Retrospectively, I am glad I took up the challenge. Far from what I feared, the insights gained from my participation in the Summer Abroad and the International Course Module programs challenged my long-held experience and compelled me to make major changes in my on-campus teaching and student learning techniques. If the unconventional, informal teaching style, content, and context of the Summer Abroad and International Course Module programs have taught me anything, it is that students' and the instructor's open, collaborative, imaginative interactions and relationships with the course material, with each other, and with the field setting lead to transformative learning (Wink 2005). Experience has shown that the transformative power of field schools is hidden in their indefinable elements that are often unrecognized in the on-campus learning environment (Mitussis and Sheehan 2013).

Field schools take place in special conditions. Those conditions are quite different from on-campus teaching and learning but, as demonstrated in this chapter, some of the magic of field schools can be harnessed to productive ends for on-campus teaching and learning. I totally concur with Brewer and Cunningham (2009), Mezirow (1997), Mezirow and Associates (2000), Difruscio and Rennick (2013), and others who have argued that the catalyst for transformative change is "disequilibrium," "disorientation," and "cognitive negotiation." Attentive to this, and within a constrained environment,[15] it has been possible to put into action and enjoy some of the benefits of the field school in a regular on-campus classroom. Adjustments have not been applied in one go, but rather incrementally and are dependent on a number of other contingencies such as class size. In an age of superfluous information, a pedagogical model that permits student immersion in their learning liberates and transforms both the student and the instructor.

NOTES

I am grateful to Elizabeth Vibert for inviting me to be part of this book project exercise and for her constructive comments on my first draft. I acknowledge my International Course Module and Summer Abroad program students – for teaching me how to experience teaching and learning differently. To my on-campus University of Toronto (2013–16) students, your eagerness for a dialogic model of teaching and learning has made me a true believer that change is possible. To Kelly Stewart (formerly of Northwestern University), thank you for running your Summer Aboard course through the Makerere Institute of Social Research during the time I was the institute's director. I gained a lot from our collaborations.

1 By this, I mean approaching class and its activities with a sense of openness, excitement, readiness to learn something about ourselves, and a willingness to engage a world around us.
2 I acknowledge and give credit to Elizabeth King who designed this course and for all the tips she gave me. I am also indebted to Yvette Ali for her vote of confidence in me that I could run the course at such a short notice.
3 Service-learning activities have included contributing to the digging of the foundations for a boarding girls' hostel and teachers' houses, planting trees, fetching water for the "Mamas," and spending half a day conversing and playing with schoolchildren.

4 This is a term used by "Me to We" to refer to Maasai mothers involved in the organization's artisan, water, and sanitation programs.

5 The two students gained appreciation of each other's initial outburst. They later became friends before we left Kenya.

6 Unashamedly, her supervisor directly communicated this to me. True, on our last day in Uganda, the teaching assistant like two other students in the group had developed an upset stomach after eating a hamburger in one of Kampala's upscale malls.

7 For example, students graded each other's work, and I often sat back and listened to them take positions on issues.

8 For a good discussion of the complexities of being in this role, see Tennant (2006, 130); Ellsworth (1992); Che, Spearman, and Manizade (2009).

9 They submit their work with a statement of collaboration, which is signed by all members of the group.

10 The Albertina and Walter Sisulu biography is 656 pages long.

11 For example, in one of the 300-level courses, in 500-word papers, the first and second assignments deal with concepts and a research proposal, followed by a less than five-page annotated bibliography paper and theoretical framework, and lastly, a final paper.

12 I attribute this concern to the fact that these students do realize that the all-around goal of these classes goes beyond assignment grades. Above all, they are now cognizant that group performance (how they work in groups) supports collaborative learning.

13 For example, I would ask: "How does today's class relate to what you already knew or plan on doing in the future?"

14 One such reason is a rejection of ideas that fail to fit their preconceptions.

15 All the classes in which I have attempted to introduce these changes officially met once a week for a two-hour session.

REFERENCES

Abeysekera, Lakmal, and Phillip Dawson. 2015. "Motivation and Cognitive Load in the Flipped Classroom: Definition, Rationale and a Call for Research." *Higher Education Research & Development* 34 (1): 1–14. https://doi.org/10.1080/07294360.2014.934336.

Alexander, Robin. 2006a. *Education as Dialogue: Moral and Pedagogical Choices for a Runaway World.* Hong Kong: Institute of Education with Dialogos.

Alexander, Robin. 2006b. *Towards Dialogic Teaching: Rethinking Classroom Talk.* 3rd ed. York: Dialogos.

Barkley, Elizabeth F. 2009. *Student Engagement Techniques: A Handbook for College Faculty.* San Francisco: Jossey-Bass.

Biggs, John, and Catherine Tang. 2007. *Teaching for Quality Learning at University.* 3rd ed. Buckingham: Open University Press.

Brewer, Elizabeth, and Kiran Cunningham, eds. 2009. *Integrating Study Abroad into the Curriculum: Theory and Practice Across the Disciplines.* Sterling, VA: Stylus.

Brookfield, Stephen D., and Stephen Preskill. 1999. *Discussion as a Way of Teaching: Tools and Techniques for University Teachers.* Buckingham: Open University Press.

Bruner, Jerome S. 1996. *The Culture of Education.* Cambridge, MA: Harvard University Press.

Buckingham, David. 2003. "Media Education and the End of the Critical Consumer." *Harvard Educational Review* 73 (3): 309–27. https://doi.org/10.17763/haer.73.3.c149w3g81t381p67.

Burbules, Nicholas C. 2000. "The Limits of Dialogue as a Critical Pedagogy." In *Revolutionary Pedagogies: Cultural Politics, Instituting Education, and the Discourse of Theory,* ed. Peter Pericles Trifonas, 251–73. New York: Routledge.

Che, S. Megan, Mindy Spearman, and Agida Manizade. 2009. "Constructive Disequilibrium: Cognitive and Emotional Development through Dissonant Experiences in Less Familiar Destinations." In *The Handbook of Practice and Research in Study Abroad: Higher Education and the Quest for Global Citizenship,* ed. Ross Lewin, 99–116. New York: Routledge.

Difruscio, Cathleen, and Joanne Benham Rennick. 2013. "Culture Shock, Cognitive Dissonance, or Cognitive Negotiation? Terms Matter in International Service Learning Programs." In *The World Is My Classroom: International Learning and Canadian Higher Education,* ed. Joanne Benham Rennick and Michel Desjardins, 63–84. Toronto: University of Toronto Press. https://doi.org/10.3138/9781442669079-006.

Ellsworth, Elizabeth. 1992. "Why Doesn't This Feel Empowering? Working through Repressive Myths of Critical Pedagogy." In *Feminisms and Critical Pedagogy,* ed. Carmen Luke and Jennifer Gore, 90–119. New York: Routledge.

Epprecht, Marc. 2004. "Work-Study Abroad Courses in International Development Studies: Some Ethical and Pedagogical Issues." *Canadian Journal of Development Studies* 25 (4): 687–706. https://doi.org/10.1080/02255189.2004.9669009.

Fielding, Michael. 2001. "Students as Radical Agents of Change." *Journal of Educational Change* 2 (2): 123–41. https://doi.org/10.1023/A:1017949213447.

Freire, Paulo. 1970. *Pedagogy of the Oppressed.* New York: Continuum.

Gokhale, Anuradha A. 1995. "Collaborative Learning Enhances Critical Thinking." *Journal of Technology Education* 7 (1): 22–30. https://doi.org/10.21061/jte.v7i1.a.2.

Kiely, Richard. 2004. "A Chameleon with a Complex: Searching for Transformation in International Service-Learning." *Michigan Journal of Community Service Learning* 10 (2): 5–20.

Kiely, Richard. 2005. "A Transformative Learning Model for Service-Learning: A Longitudinal Case Study." *Michigan Journal of Community Service Learning* 12 (1): 5–22.

Kolb, Alice Y., and David A. Kolb. 2009. "Experiential Learning Theory: A Dynamic, Holistic Approach to Management Learning, Education and Development." In *The Sage Handbook of Management Learning, Education and Development*, ed. Steven J. Armstrong and Cynthia V. Fukami, 42–68. Los Angeles, CA: Sage. https://doi.org/10.4135/9780857021038.n3.

Lipman, Matthew. 2003. *Thinking in Education*. 2nd ed. Cambridge: Cambridge University Press. https://doi.org/10.1017/CBO9780511840272.

McCaleb, Sudia Paloma. 1998. "Connecting Pre-Service Teacher Education to Diverse Communities: A Focus on Family Literacy." *Theory into Practice* 37 (2): 148–54. https://doi.org/10.1080/00405849809543798.

McKeachie, Wilbert J., Paul R. Pintrich, Yi-Guang Lin, David A.F. Smith, and Rajeev Sharma. 1990. *Teaching and Learning in the College Classroom: A Review of the Research Literature*. 2nd ed. Ann Arbor, MI: National Center for Research to Improve Post-Secondary Teaching and Learning.

Mezirow, Jack. 1997. "Transformative Learning: Theory to Practice." *New Directions for Adult and Continuing Education* 74: 5–12. https://doi.org/10.1002/ace.7401.

Mezirow, Jack, and Associates. 2000. *Learning as Transformation: Critical Perspectives on a Theory in Progress*. San Francisco: Jossey-Bass.

Mitussis, Darryn, and Jackie Sheehan. 2013. "Reflections on the Pedagogy of International Field-schools: Experiential Learning and Emotional Engagement." *Enhancing Learning in the Social Sciences* 5 (3): 41–54. https://doi.org/10.11120/elss.2013.00013.

Mok, Heng Ngee. 2014. "Teaching Tip: The Flipped Classroom." *Journal of Information Systems Education* 25 (1): 7–11.

Mycroft, Lesley. 2009. "The Reflective Learning of Humanities Students: Is Their Learning Enhanced by a Blended Approach?" *Enhancing Learning in the Social Sciences* 2 (2): 1–25. https://doi.org/10.11120/elss.2009.02020004.

Rennick, Joanne Benham, and Michel Desjardins, eds. 2013. *The World Is My Classroom: International Learning and Canadian Higher Education*. Toronto: University of Toronto Press. https://doi.org/10.3138/9781442669079.

Shor, Ira. 1992. *Empowering Education: Critical Teaching for Social Change*. Chicago: University of Chicago Press.

Shor, Ira. 1996. *When Students Have Power: Negotiating Authority in a Critical Pedagogy*. Chicago: University of Chicago Press.

Sichel, Benjamin. 2006. "'I've come to help': Can Tourism and Altruism Mix?" *Briarpatch Magazine*, 2 Nov. Accessed 25 Feb. 2016. https://briarpatchmagazine .com/articles/view/ive-come-to-help-can-tourism-and-altruism-mix/.

Slavich, George M., and Philip G. Zimbardo. 2012. "Transformational Teaching: Theoretical Underpinnings, Basic Principles, and Core Methods." *Educational Psychology Review* 24 (4): 569–608. https://doi.org/10.1007/s10648-012-9199-6.

Smith, Elizabeth. 2011. "Teaching Critical Reflection." *Teaching in Higher Education* 16 (2): 211–23. https://doi.org/10.1080/13562517.2010.515022.

Strayer, Jeremy F. 2012. "How Learning in an Inverted Classroom Influences Cooperation, Innovation and Task Orientation." *Learning Environments Research* 15 (2): 171–93. https://doi.org/10.1007/s10984-012-9108-4.

Tennant, Mark. 2006. *Psychology and Adult Learning.* 3rd ed. Abingdon: Routledge.

Wegerif, Rupert. 2007. *Dialogic Education and Technology: Expanding the Space of Learning.* New York: Springer. https://doi.org/10.1007/978-0-387-71142-3.

Wegerif, Rupert. 2008. "Dialogic or Dialectic? The Significance of Ontological Assumptions in Research on Educational Dialogue." *British Educational Research Journal* 34 (3): 347–61. https://doi.org/10.1080/01411920701532228.

Wink, Joan. 2005. *Critical Pedagogy: Notes from the Real World.* 3rd ed. Boston, MA: Allyn and Bacon.

THERE IS NO FRONT OF THE CLASSROOM HERE ...

After completing my first two years of my undergraduate degree in Earth and Ocean Sciences, I had fallen into the naive opinion that I was beginning to know the material very well. When I enrolled that summer in a field-based course, I realized after the first few days that, although I may be competent at the basics, the world of geology and earth science was a vast field of study, and so far I had only scratched the surface. The first day of the summer course I arrived on campus and was greeted by the familiar faces of students I had taken classes with ... many of whom I didn't actually know by name. I was flustered, nervous, and extremely curious about how things were going to go. The next three weeks would be spent in the field with this group of people, and I honestly had no idea what to expect.

All of the instructors, including the course professor, the lab instructors, and the teaching assistants, helped to make the course manageable. The material covered was in-depth, hands-on, and strenuous. Our time was spent hiking through the bush and mapping the geological terrain in pairs or teams. At the conclusion of the course many things had changed for me. I became comfortable talking and conversing with all of the instructors, who had seemed intimidating at the beginning. My group of friends had expanded to include many of those nameless faces that were only vaguely familiar at the beginning of the course. Many of these friendships would continue on throughout the remainder of my degree and still continue today.

My knowledge of geology and what it meant to be a geologist skyrocketed. Fieldwork helped me to grow in more ways than just academically. Through the field course I grew to understand more about "real-world" interactions instead of just teacher-student and student-student interactions. I grew more confident in myself and my abilities in a way that would never have been possible through traditional classroom-based learning. Being in the field created a tangible link to the real world, allowing me to truly understand why we were learning the material we were learning. The other valuable part of learning in the field is that there is no front of the classroom, no assigned seating or rows, no single form of presentation; everything is tangible, visible, and right in front of you. Learning in the field allowed me to learn in ways that were better suited to my learning style and promoted my growth as a student. In the entirety of my degree I never once took another course that changed me, taught me, and inspired me as much as this field-based course. Learning in the field allowed me to grow into more than just a student, it allowed me to grow into a functioning member of the community that I was working towards joining.

Rob Cook
Participant, University of Victoria's
Earth Science Field School

4

Settlers Unsettled: Using Field Schools and Digital Stories to Transform Geographies of Ignorance about Indigenous Peoples in Canada

HEATHER CASTLEDEN, KILEY DALEY,
VANESSA SLOAN MORGAN, AND PAUL SYLVESTRE

Introduction

Indigenous-Settler Relations

Globally, Indigenous peoples have been subjected to subordinating colonial policies for generations. Implemented by imperial governments, these policies have largely been designed to silence Indigenous voices and delegitimize their cultures and governance structures in place of those of the settler[1] (largely white) populations (Louis 2007). It has been well documented that geographical practices have played a significant role in the colonial enterprise (Godlewska and Smith 1994; Powell 2008; Painter and Jeffrey 2009): Many Indigenous peoples have been removed from their traditional territories through allotment and physical dispossession, for example, from reserve/reservation systems (Bracken 1997; Harris 2002); most Indigenous territories have been reinscribed with European-defined political borders (Alfred 2005; Simpson 2008); Indigenous place names have, for the most part, been replaced with names from European homelands or explorers, for example, the Canadian provinces of British Columbia, New Brunswick, and Nova Scotia were named in reference to the United Kingdom (Heikkila 2007; Simpson 2008); and colonial governments have claimed sovereignty over Indigenous territories that were, in many cases, unceded. The social construction of political borders pays tribute to the continued role colonialism occupies in defining the geopolitical and sociopolitical landscape; its effects and philosophies are not, however, a thing of the past.

Within the Canadian context, non-Indigenous peoples' lack of awareness of and misinformation about Indigenous world views and lived colonial experiences – e.g., residential schools, the Indian Act, enfranchise-
ment, criminalization of spiritual practices, etc. – is influenced by their
systematic exclusion from educational curricula; thus, as Godlewska and
colleagues (2010)[2] have suggested, a "geography of ignorance" of Indigenous realities is produced and perpetuated. The question of how to overcome this ignorance, which contributes to colonial mentalities and racist
attitudes, in post-secondary education that provides at least a cursory
introduction to understanding Indigenous ontologies and epistemologies, as well as the colonial processes that impact Indigenous-settler relations, is an ongoing challenge (Turner 2006; Warry 2007). This chapter
seeks to answer that question by exploring the use of an arts-based pedagogical tool – digital storytelling – as a means of assessing non-Indigenous
students' perspectives of the transformative value of bringing them out
from behind books and lecture halls to engage in a community-based
Indigenous Perspectives on Resource and Environmental Management
field school that allowed them to have direct interaction with and learn
from Indigenous peoples in Indigenous spaces and places.

The Role of Geography Education

Unquestionably, the subject of geography attracts students with diverse
scholarly interests and as a discipline that encompasses physical, social,
and health sciences, applied practice, and theory building, geography
educators often struggle with how to integrate multiple perspectives into
educational approaches. In Canada, for example, geography's emerging
engagement with Indigenous Knowledge holders[3] is growing (Castleden,
Mulrennan, and Godlewska 2012), yet the traditional classroom limits
opportunities to interact with those who hold alternative world views
(Pløger 2001), especially when universities are still very much complicit
in a colonial mentality (Mihesuah and Wilson 2004). For university professors and students trained in the post-positivist tradition, the tendency
is to (often unwittingly) appropriate the knowledge of the "other" (Said
1978) and assimilate it into the greater whole. The reality, though, is that
different ways of knowing are not reducible to each other (Van Eijck
and Roth 2007) and to do so is not necessarily engagement, it is colonial
(Battiste and Youngblood Henderson 2000). While attempts are made to
engage the "other," the conventional pedagogy of knowledge transfer in
the academy tends to occur: expert knowledge flows from professors to

students (Fletcher and Cambre 2009). Such an approach is poorly positioned, in an epistemological sense, to foster the sort of individual transformation required for engagement that transcends tokenism (Kollmuss
and Agyeman 2002; Moore 2005; Wright 2006).

Nowhere in geography is a deeper level of understanding needed
more than in the sub-discipline of resource and environmental management, especially given Indigenous peoples' relationships to the land
(Parkes 2011) and the ongoing "Indian land question" in the context
of modern treaties and land claim settlements in Canada (Harris 2002).
Graduates of such programs, who are still for the most part non-Indigenous
people, wanting to work in this field must leave their programs with a
solid understanding of treaty and Aboriginal rights and title not only
from a Western perspective, but also from Indigenous perspectives, as
they will encounter these issues with growing frequency given the rising
political power of Indigenous peoples in Canada. The problem is that
direct engagement with Indigenous peoples – in Indigenous spaces and
places – to understand their (multiple) situated perspectives often does
not occur. Despite the inclusion of Indigenous voices vis-à-vis assigned
readings or guest lectures in environmental management courses, a dis-
connect between the information transferred in the university and the
professional practices of environmental managers exists (Louis 2007).
With little to no experience or understanding of the historically oppressive realities that have impacted Indigenous peoples and their continued
exclusion from governance and decision-making conversations, many,
if not most, graduates lack awareness of the contemporary (often contentious) relationships that exist between Indigenous and settler populations and governing structures. For example, while the current Idle
No More movement in Canada, an Indigenous-led peaceful expression
of resistance to changes in federal policies that will weaken environmental protection and treaty rights, is gathering momentum as well as
non-Indigenous supporters both within the nation and beyond, it is also
bringing out, with great clarity and even greater numbers, sentiments of
deep-rooted racism towards Indigenous peoples across the country.

Transformative Learning Approach

Engagement in experiential education that leads to transformative learning (Mezirow 2009) offers a potential solution to the challenge described
above. Transformative learning is an educational theory that seeks to
promote "a critical dimension of learning ... that enables us to recognize,

reassess, and modify the structures of assumptions and expectations that frame our tacit points of view and influence our thinking, beliefs, attitudes, and actions" (Mezirow 2009, 18). It allows students to conceptualize and actively immerse themselves in a process of exploring their subjectivity and positionality in relation to subject matter while simultaneously challenging students to actively engage with concepts that may be unfamiliar and challenge their existing beliefs. As a philosophy, it primarily draws on the emancipatory educational ideology of Paulo Freire as well as the critical theories of knowledge and communication advanced by Jürgen Habermas (Moore 2005; Mezirow 2009). Individuals begin to understand themselves as socially constituted and socially constitutive agents, where they come to understand how their world view is culturally contingent as well as coming to understand that as historically situated agents they play an active role in reconstituting the socially constructed world (Kreber and Cranton 2000) and undergo cognitive and affective transformation.

To provide a transformative learning experience (and assess whether it has taken place), educators use a variety of techniques including, for example, artistic expression, experiential learning, and journal writing (Cranton 2002; Lawrence 2008). Teachers and students often describe this process as disorienting, unsettling, and deeply personal in nature (Moore 2005). This chapter, as stated earlier, explores the use of an arts-based education tool – digital storytelling – as a means of assessing transformative learning in a community-based field school that brings (mainly) non-Indigenous students out of the classroom and into Indigenous communities to learn directly from Indigenous Knowledge holders. Digital storytelling is a multimedia-based, multidimensional means of communication that integrates the vision and voice of the storymaker (Burgess 2006; Lambert 2008). By actively engaging students throughout the storymaking process, digital storytelling was, in this case like in many others using the technique, intended to encourage students to be reflexive with the course content (Fletcher and Cambre 2009). Worth noting, it is also well aligned with Indigenous oral traditions (King 2003; Cherubini 2008; Cunsolo Willox et al. 2013).

Research Context and Methods

ENVI5039: Indigenous Perspectives on Resource and Environmental Management is a graduate-level elective field school offered in the School for Resource and Environmental Studies at Dalhousie University in Halifax, Nova Scotia, Canada. The course involves a two-day orien-tation followed by a week-long community-based field school compo-nent. Co-facilitated by a non-Indigenous professor and an Indigenous community member, students (in this case, all were non-Indigenous) are introduced to several rural and semi-urban Indigenous (Mi'kmaq) communities and organizations in Nova Scotia where they engage in cultural and natural resource-based activities (e.g., ceremony, sharing circles, medicine walks, eel fishing, etc.) and discuss resource issues with Mi'kmaq Knowledge holders, leaders, and Elders. A weekly class-room-based component follows where students continue to learn about Indigenous world views through guest lectures with Indigenous people, assigned readings, and associated assignments. In 2010, digital story-telling was one of several tools used to assess student learning. Students completed a three-day workshop wherein they created digital stories[4] about what they learned.

In preparation for the workshop, students were expected to keep a journal throughout the duration of the field school; they were encour-aged to return to their writing throughout the experience to reflect on their (potentially changing) positionality and perspectives on the course themes. A week prior to the workshop, they were briefly introduced to examples of completed digital stories created in other venues for other purposes. They were asked to prepare the following: a one-page story script of their personal reflections on what they learned (approximately 500–750 words); compile ten to thirty images or video clips that related to the story they wanted to create, which could have been taken dur-ing the field school or sourced from the Internet; acquire one to three music pieces (ideally instrumental only) that might "fit" with the themes or goals of their stories; and for those who had them, bring MacBook computers (those who did not have MacBooks were provided with one for the duration of the workshop).

On Day 1 of the workshop, students were provided with an overview of the goals of digital storytelling, drawing from the materials provided in Joe Lambert's (2010) *Digital Storytelling Cookbook*. The remainder of the day was spent conducting a story circle, whereby participants shared their story (script), talked about the pictures and music they had selected, and received feedback and constructive suggestions for their story develop-ment from their peers and professor;[5] they also viewed iMovie tutorials (the software on a MacBook that was used to create digital stories), and began storyboarding (matching their script with the selected images, visual and audio transitions, and music).

On Day 2, students were provided with a thirty-minute live demonstration of the iMovie software wherein the professor created a short digital story, and students gained an understanding of how to make their stories; students were then given time and space (and technical support) to finalize their scripts and record them, after which they began importing audio and visual materials and "rough" editing their stories. On Day 3, students made final edits to their digital stories through the use of transitions (still and moving visuals and music fading in and out) to allow them to improve the flow and impact of their stories. A private celebratory screening among the students and professor was held at the conclusion of the workshop.

The digital stories, which were between three and five minutes long, were then screened at a public event involving the university community and members of Mi'kmaq communities and organizations across the province. The stories were powerful and profound, and prompted the professor, a researcher, to explore the transformative potential of digital storytelling as a mechanism for contributing to ways of overcoming a "geography of ignorance" among future non-Indigenous environmental managers. She established a research team to pursue this line of inquiry, and six months after the course was completed, the students, all non-Indigenous ($n = 8$),[6] were contacted via email and agreed to participate in a qualitative semi-structured interview and short questionnaire for this study.[7] One member of the research team conducted all interviews to ensure continuity in the interview process (Gubrium and Holstein 2003). All interviews were transcribed verbatim and returned to participants to confirm accuracy (Baxter and Eyles 1997). During interviews, participants also completed a written questionnaire, which was adapted from Ramsden's (1991) course experience survey. It was used to encourage participants to reflect on their learning experience and the process of making digital stories. In addition to these data, participants gave permission to the research team to use each participant's digital story as well as their pre-course objectives form, which the instructor used to identify and understand student motivations for enrolling in the course. Data were analysed inductively (Glaser 1992) through the collective development of a coding scheme (Gubrium and Holstein 2003). Two team members completed the coding and their analyses were compared to ensure consistency (Baxter and Eyles 1997; Whittemore, Chase, and Mandle 2001). The research team completed a thematic analysis, involving categorization of coded data and a comparison of those categories with existing literature (Aronson 1994). Three key themes – (1) openness to transformation, (2) transformation through relationships, and (3) vulnerability and pride – emerged from the data. These themes are used to organize the following section of this chapter.

Findings

Openness to Transformation

All participants expressed awareness of the importance of including Indigenous voices in environmental management decision-making. However, they felt their earlier education did not provide them with a rich understanding of issues. They came to the course with an eagerness to fill that gap and an open willingness to the learning experience being offered:

> We don't learn anything about [Indigenous perspectives] in school and then everything that we know is from either what we hear in the media, what we read for ourselves, what we hear people talking [about] around us ... I wanted to learn more about where it was coming from. (Anonymous student, pers. comm.)

With most participants having never knowingly[8] engaged with Indigenous peoples in Indigenous places and spaces (never having, for example, visited a rural Indian reserve or an urban Friendship Centre), one participant brought up the notion of the "museum-ized Indian" to highlight the static means by which Indigenous communities are often imagined: "they're ... not wearing loincloths" (pers. comm.). A different participant indicated that their initial intent for enrolling in the class was for professional reasons but that during the course their motivation shifted to:

> a personal quest ... personal interest ... A general new world view [of] being more open minded and understanding other ways of living not just in the same province, in the same country. (Pers. comm.)

Still another, reflecting on the structurally oppressive colonial mentality and institutional framework that continues to exist, stated:

> I did not know ... residential schools were still open until 1996. That is still fairly recent and just little facts about those kind[s of things] shifted my

way of [thinking] ... not looking at Indigenous communities but looking at society, Canadian society in general. (Anonymous student, pers. comm.)

Whether depicting segregated knowledge(s) through metaphors of silos, or flashing historical representations to communicate the importance of questioning the location from which concepts of truth originate, each digital story located the participant within the subject matter. These epiphanies were positioned in relation to the course material:

What I tried to do was to apply the learning of my personal experience and see how that fit. And that was a way ... to internalize what the course had brought me ... It was all about the experience. (Anonymous student, pers. comm.)

Each digital story involved personal accounts of participants' acknowledgment of their limited and preconceived notions of Indigenous peoples and perspectives. Their stories can be seen as a commentary not only on the story-makers themselves, but also on the inadequacies of their earlier education to address these issues.

Transformation through Relationships

One of the ways to overcome a "geography of ignorance" is relationship building. In the pre-course objectives, all participants expressed an interest in learning about the non-Indigenous professor's lived experience in working with Indigenous peoples. They wanted someone, as one participant put it,

with practical experience in the field, and not just an understanding that can be acquired from textbooks. (Pers. comm.)

With that in mind, the professor emphasized her role as a facilitator, not a teacher in the traditional sense, and participants echoed the sentiment that while it was frustrating at times to not be told what to do and how to do it, they valued the fact that she "acted more as a guide in the learning experience" (Anonymous student, pers. comm.).

The community-based component also led to a sense of community within the group. Sharing circles,[9] for example, encouraged students to communicate personal thoughts, not always an easy task:

This was an emotional journey ... and it was a learning experience ... and ...
that is hard to communicate especially with this day and age or our generation anyways. We are all about keeping feelings and that kind of stuff within
us. (Anonymous student, pers. comm.)

Moving beyond intellectual engagement to experiencing emotion,
whether positive or negative, is considered key to the transformative learning process (Lawrence 2008). The field school component
placed students, half of whom had no prior contact with each other,
in unfamiliar environments with unfamiliar people. One participant
stated:

I think if you had the wrong group of people in a room it would have been ...
uncomfortable. (Pers. comm.)

However, another participant discussed how the process of community
building prepared them for including personal elements in their digital
story and sharing it with the class:

I was a little more comfortable ... discussing ... personal feelings in talking
about things that you don't typically talk about with people at school ...
Sharing circles ... [were] a very good bonding process so afterwards I feel
like I probably felt less vulnerable then [*sic*] if I had just [shown my digital
stories] to strangers. (Pers. comm.)

Although the digital story was an individual assignment, all participants mentioned the importance of their new interpersonal relationships while making the stories. During the script-writing process of the
three-day workshop, for example, one participant recalled:

Some people really struggled [with] writing their stories and [started] getting really stressed over it so that was not fun because whenever anybody's
stressed, everybody's stressed, especially in a small group. (Pers. comm.)

Participants stated that making their stories allowed them to share
their journey "versus reading a whole bunch of papers and synthesiz[ing]
what others have written" (Anonymous student, pers. comm.). Personalized experiences were shared rather than, as one participant stated:
"regurgitating facts" (pers. comm.). Another participant recognized
that "Mi'kmaq people know their history"; the digital stories were not

intended to tell that history again. They were, however, a way of "demon-strating to [community teachers] that [we] did gain something" (pers. comm.). In fact, participants felt they had formed relationships with those Mi'kmaq Knowledge holders they had interacted with, many of whom had shared personal stories, and there was a need among the par-ticipants to reciprocate. As one participant noted, digital stories were a "different" way to share what they learned, and another quipped:

> Who is going to read your final paper but your professor? So what? [A] digital story is … really easy to share. (Pers. comm.)

Another participant commented:

> [Digital stories] show that what we had learned at each of [the communities/ organizations] was kind of put into our experiences and put back into an oral way of talking and sharing and exchanging it with them because they had exchanged so [many] stories … I think [sharing our stories with them] was the best part of the whole course. (Pers. comm.)

Students used digital stories to communicate their experience and to relay their gratitude to community teachers, without whom the transfor-mative process would not have been possible.

Vulnerability and Pride

The process of confronting their knowledge gaps regarding Indigenous perspectives and then sharing that experience through digital stories in a public arena evoked feelings of both vulnerability and pride among the participants. Participants reflected on the personal nature of the story-making process: "It was my own thing instead of … a reproduction of facts," stated one participant, acknowledging that digital story mak-ing inherently involves "sharing something about yourself. The duality of such an affective experience was likened to "revealing a whole other side of you" (Anonymous student, pers. comm.). Another participant commented:

> I held myself to higher expectations [to] a group of people who have helped you through this process. You want to own what you are making. (Pers. comm.)

The transformation from vulnerability to pride over the course of the digital story-making process was palpable in the interviews. One participant explained:

> I was really scared because it was really personal. It was not like a group project, I couldn't hide under a group name. After we presented it, I was really proud of what I did. It was challenging but really rewarding to see the results. I just wanted to show everyone. (Pers. comm.)

The course goal, to share stories with community teachers, also evoked a sense of unease: "I did not want to sound naive" (anonymous student, pers. comm.). Despite this, participants were proud of their stories' key messages. They also recognized that peer mentorship during the story-making workshop allowed participants to "clarify what [they] needed to include" (anonymous student, pers. comm.). In short, they did not want to convey ignorance or expertise about something they were just beginning to understand. Creating individual stories in a workshop environment allowed them to draw on their peers and their professor to check their understanding.

Discussion

As a pedagogical tool, digital storytelling was an important part of *beginning* the transformative learning process for students in a community-based field school on Indigenous Perspectives on Resource and Environmental Management. The transformation process – invoking themes of openness or "readiness" to learn, the importance of relationships, and the disorienting and unsettling feelings of vulnerability and pride – could not have been possible without the community-based component, sharing circles, continuous reflective journalling, or the creation of spaces in which students felt safe enough to share personal aspects of their learning experience.

Initial discussion of openness, centring on motivations for taking the course, takes the form of a willing admission of ignorance on the part of these students about their knowledge of Indigenous issues and Indigenous perspectives of resource and environmental management. This cohort sought to earnestly engage with the course material to actively address their academic and professional shortcomings; the recognition of significant knowledge gaps was the necessary impetus to begin the transformative learning process. By consciously identifying gaps in their education, these students demonstrated receptivity to a non-conventional educational opportunity. Without this, the goals of the course – recognition of Indigenous values in resource and environmental management – would have been difficult to attain. A "disorienting dilemma" (Mezirow 2009), the first step in the transformative learning process, is what prompted these students to begin critically reflecting on certain personally held preconceptions.

The transformation process continued once community engagement was underway. For five of the seven students this was the first time they had been in an Indigenous (reserve) community (one of many Indigenous spaces and places). This experience was cited as the most important component of the course, with students suddenly finding themselves immersed in destabilizing (but positive) situations with the "other." Direct engagement with Indigenous peoples was an opportunity for them to be exposed to unfamiliar world views. Much of the inspiration for what students would later include in their stories was gathered during this section of the course. That students' stories tended to situate themselves in the context of this experience speaks to critical reflection on their part, an instrumental aspect of the transformation process (Moore 2005; Mezirow 2009). Most participants highlighted the importance of the relationships forged during this period. The emotions, at times tumultuous, as students identified their own biases and beliefs, which are associated with transformative learning, are a direct result of the vulnerability required for this process (Moore 2005). Thus, creating safe and supportive spaces is integral to transformative learning. This is evidenced by the students' discourse about the camaraderie and relationships forged throughout the course. The bond created by the shared experience of students in this particular case was an important hedge against overburdening emotions that such vulnerability can entail.

Equally important were the relationships established with both the university-based and community-based teachers. In speaking about their learning experiences during the interviews, students made a point of referring to their co-instructors/facilitators and to the community teachers – although not considered experts in the traditional academic sense (i.e., none held PhDs) – with deep appreciation and feelings of connectedness with them. These distinctions speak to a more even distribution of power than that which is generally experienced in conventional classrooms. In this we observe how transformative learning theory draws on the emancipatory potential within critical pedagogical theories; see Freire (1971) and Mezirow (2009), for example. An even distribution of power in the classroom sees teaching scholarship transition from a "teacher-to-student" knowledge transfer linearity into a dialogic process (Freire 1971). Student empowerment is more effective vis-à-vis critical reflection compared with that of rote memorization, which focuses only on the students' ability to regurgitate facts. Within this entire transformational educational project, the digital storytelling component provided a necessary space for that critical reflection to happen. Furthermore, the student-teacher relationships that grew throughout this process led to a strong desire on the students' part to share their digital stories with their (community-based) teachers, with expressed sentiments among the cohort the likes of "I can't wait to share this with them." This need for reciprocity also suggests efficacy in the transformative process and the effectiveness of digital storytelling as a communicative assessment tool, one that moves beyond conventional assessment of the linear transfer of information.

The course culminated with a public screening of the digital stories. The safe and supportive spaces that had been built throughout the course created a setting where students felt comfortable sharing their stories with one another and with a wider audience (approximately 100 people attended from the academic community, Aboriginal organizations, and the general public). The positive and supportive feedback shared between peers created a sense of pride among the story-makers and the accomplishments of their peers. The stories were also an opportunity to publicly demonstrate how the course as a whole had affected them, simultaneously acknowledging their own vulnerability concerning the deeply personal nature of their stories and call attention to their own settler ignorance of the lived experiences of Indigenous peoples in Canada. These feelings, which are typically associated with acts of creative expression, are key to the rationale for the use of arts-based approaches in the transformative learning process (Lawrence 2008). Sharing personal stories in a public place, whether through theatre or film, is a powerful means of demonstrating the existence of multiple ways of knowing (Butterwick and Lawrence 2009). Within the course context, using art as a learning tool to uncover multiple ways of knowing embodies one of the central goals of the course: to introduce students to Indigenous perspectives. Through using and adapting Indigenous oral traditions, course themes and content are reflected while enabling students to explore their learning experience. Moreover, this broadening of perspective parallels one of the key goals of transformative learning: openness to alternative points of view (Mezirow 2009). Thus, settlers were unsettled and from there they were beginning to better understand themselves and their ancestors in relation to the "other" and came to a new place of knowing. This is not to say that their transformational process of "unlearning"[10] was over, not by a long stretch. The digital stories and the field school as a whole were the first steps in a lifelong journey of coming to understand our historical, present, and future relationships with one another.

Conclusion

Digital storytelling as an arts-based assessment tool transcends the normative confines of conventional classroom assignments (Lawrence 2008). In this context, a graduate-level field school on Indigenous perspectives of resource and environmental management in conjunction with digital stories effectively served multiple roles. First, from the perspective of transformative learning, the field school enabled students to emotionally, mentally, physically, and spiritually engage with Indigenous (Mi'kmaq) peoples about the key course concepts and to truly immerse themselves in the experiential learning process. Second, because digital storytelling is a tangible product that individuals produce, it meets the requirements of the academic institution to evaluate students' progress. Assessment of transformative learning is often incompatible with conventional means of evaluation. Digital storytelling, however, with its emancipatory potential, offers an alternative to formal epistemologies, language, and academic traditions, which are largely patriarchal or colonial in origin (Lawrence 2008). The act of telling stories about experiences and the actual sharing of these stories creates a platform for story makers and viewers to "reconsider the meanings of their experiences" (Butterwick and Lawrence 2009, 35). Third, and finally, in a course dedicated to Indigenous perspectives, digital storytelling and the act of sharing is reflective of Indigenous oral traditions (King 2003; Cherubini 2008; Cunsolo Willox et al. 2013). Through both of these aspects of the learning process, the existence of multiple ways of knowing became evident. A field school and digital storytelling created the space necessary for students to acknowledge their settler identity, grapple with this identity, situate themselves in their *ongoing* learning journey with overcoming their ignorance concerning Indigenous peoples' multiple realities in Canada, and document their own personal and professional transformations.

There is a risk, albeit small, that those who have begun such a journey may declare themselves "transformed" (read: know enough), but in this case, students were continually exposed to the philosophy on a lifelong journey; notwithstanding this, we do end with a caveat: that completion of one field school on Indigenous perspectives of resource and environmental management is, by no means, "enough" in terms of transforming our awareness and understanding. Among this cohort of non-Indigenous students or any others for that matter, it is indeed a continual process of critical engagement about Indigenous-settler relationships in Canada.

Racism exists in Canada; it is widespread and ingrained across the landscape. In the current "Idle No More" climate in this country, to downplay these tensions between Indigenous peoples and settlers would be disingenuous. While this chapter is about the unsettling of non-Indigenous students participating in the study through the use of digital storytelling, this chapter may also have an unintended consequence: to unsettle other settlers, that is, to induce reflexivity in pedagogical processes and curricular content.

NOTES

This chapter was originally published as follows: Heather Castleden, Kiley Daley, Vanessa Sloan Morgan, and Paul Sylvestre (2013), "Settlers Unsettled: Using Field Schools and Digital Stories to Transform Geographies of Ignorance about Indigenous Peoples in Canada," *Journal of Geography in Higher Education* 37 (4): 487–99, and is reprinted here by permission of the Taylor & Francis Ltd., http://www.tandfonline.com, with minor editorial changes. All authors contributed equally to the research as well as the development and preparation of this manuscript; order of authorship is strictly alphabetical in nature. We wish to extend our sincere thanks to the participants who agreed to share their perspectives and their digital stories for this study. Our deep appreciation is also extended to all those with whom we interacted and learned from during the field school, from Mi'kmaq lawyers to Mi'kmaq chiefs, Mi'kmaq Elders, treaty negotiators, Traditional Knowledge holders, Mi'kmaq natural resource managers, Mi'kmaq Sweat Lodge keepers, and Mi'kmaq eel fishers – and many more. We could not do this field school without you. Wela'lin.

1 "Settler" refers to non-Indigenous peoples whose ancestors, or who themselves, have immigrated to Canada to inhabit Indigenous territories and subsequently dispossess Indigenous peoples.
2 For a rich discussion of this concept, see Godlewska and colleagues' (2010) paper; the idea of a "geography of ignorance" rather than a "history of ignorance" is used in this context because this is not an issue of the past, it is an ongoing problem plaguing Indigenous-settler relations in Canada.

3 To describe this way of knowing, Indigenous scholar Brant Castellano
   (2000) characterizes Indigenous Knowledge as oral, experiential, holistic,
   personal, and based within a storied or metaphoric language (Hart 2010).
   While succinct definitions for Indigenous Knowledge do exist in the
   literature, the choice to describe, versus define, is done so here to offset
   efforts that attempt to compare epistemologies of which there are no
   grounds for such evaluation (McGregor 2004). Indigenous scholars Battiste
   and Youngblood Henderson (2000) and Hart (2010) suggest that those
   who are not Indigenous Knowledge holders focus on grasping the process
   of understanding, rather than attempts at delineating, an exercise that
   encourages recognition of differential realties.

4 During preparation of this manuscript, permission was sought from the
   participants to post their digital stories on the first author's academic
   website. The digital stories of those who granted permission can be viewed
   at: https://www.dal.ca/faculty/management/sres.html.

5 The professor sought feedback from the students about the three-day
   workshop; they indicated a strong desire to hold the Day 1 story circle at
   least one week in advance of the workshop to give them more time to refine
   their ideas, revise their stories based on the group's feedback, and gather
   the audio and visual resources they wanted for their evolving stories. In a
   subsequent year, the professor incorporated this feedback in the course
   schedule, and students seemed much more relaxed with respect to what
   they needed to accomplish in the three-day workshop.

6 In an era of administrative "austerity" measures with respect to university
   budgets on campuses across Canada, which is prompting many to deliver
   massive open on-line courses (MOOCs), this field school consistently
   requires justification within its administrative unit to proceed given its small
   size (a minimum of six students must register for the course to proceed
   [read "break-even"] and a maximum of twelve students can enroll [read
   "manage logistically"]) and high costs, which are wholly borne by students.

7 The third author completed this course. However, given her role on the
   research team, she was excluded as a participant to limit potential bias
   (Briggs 2003); instead the interview guide was pilot-tested on her and
   subsequently revised (Lincoln and Guba 1985).

8 Indigenous peoples continue to encounter discrimination and racism in
   Canada; as a result, studies have found that some choose to hide their
   identity to avoid such encounters; see, e.g., Berry (1999), Canales (2004),
   and Castleden et al. (2010).

9 Sharing circles involve turn-taking that is often formalized with ceremony
   (e.g., smudging with sacred medicines or the use of a sacred talking stick

or stone) and hold sacred meaning for many Indigenous peoples and their allies in terms of spiritual and emotional growth (Lavallée 2009). The experience involves sharing of one's whole self, not just knowledge sharing, and those in the circle are expected to exhibit respectful listening and non-judgmental, helpful, and supportive discourse (Lavallée 2009).

10 See Anne Hickling Hudson and Peter Mayo's (2012) editorial introduction for reference to "unlearning" colonial stereotypes and ideologies.

REFERENCES

Alfred, Taiaiake. 2005. *Wasáse: Indigenous Pathways of Action and Freedom.* Peterborough, ON: Broadview Press.

Aronson, Jodi. 1994. "A Pragmatic View of Thematic Analysis." *Qualitative Report* 2 (1): 1–3. http://nsuworks.nova.edu/tqr. Accessed 24 Feb. 2017.

Battiste, Marie, and James Youngblood Henderson. 2000. *Protecting Indigenous Knowledge and Heritage: A Global Challenge.* Saskatoon: Purich.

Baxter, Jamie, and John Eyles. 1997. "Evaluating Qualitative Research in Social Geography: Establishing 'Rigour' in Interview Analysis." *Transactions of the Institute of British Geographers* 22 (4): 505–25. https://doi.org/10.1111/j.0020-2754.1997.00505.x.

Berry, John W. 1999. "Aboriginal Cultural Identity." *Canadian Journal of Native Studies* 19 (1): 1–36.

Bracken, Christopher. 1997. *The Potlach Papers: A Colonial Case History.* Chicago: University of Chicago Press.

Brant Castellano, Marlene. 2000. "Updating Aboriginal Traditions of Knowledge." *Indigenous Knowledges in Global Contexts: Multiple Readings of Our World,* ed. George J. Sefa Dei, Budd L. Hall, and Dorothy Goldin Rosenberg, 21–36. Toronto: University of Toronto Press.

Briggs, Charles L. 2003. "Interviewing, Power/Knowledge and Social Inequality." In *Postmodern Interviewing,* ed. Jaber F. Gubrium and James A. Holstein, 242–54. Thousand Oaks, CA: Sage. https://doi.org/10.4135/9781412985437.n13.

Burgess, Jean. 2006. "Hearing Ordinary Voices: Cultural Studies, Vernacular Creativity and Digital Storytelling." *Continuum* 20 (2): 201–14. https://doi.org/10.1080/10304310600641737.

Butterwick, Shauna, and Randee Lipson Lawrence. 2009. "Creating Alternative Realities: Arts-Based Approaches to Transformative Learning." In *Transformative Learning in Practice: Insights from Community, Workplace, and Higher Education,* ed. Jack Mezirow, Edward W. Taylor, and Associates, 35–55. San Francisco: Jossey-Bass.

Canales, Mary K. 2004. "Connecting to Nativeness: The Influence of Women's American Indian Identity on their Health Care Decisions." *Canadian Journal of Nursing Research* 35 (4): 18–44.

Castleden, Heather, Monica Mulrennan, and Anne Godlewska. 2012. "Community-Based Participatory Research Involving Indigenous Peoples in Canadian Geography: Progress? An Editorial Introduction." *Canadian Geographer* 56 (2): 155–9. https://doi.org/10.1111/j.1541-0064.2012.00430.x.

Castleden, Heather, Valorie A. Crooks, Neil Hanlon, and Nadine Schuurman. 2010. "Providers' Perceptions of Aboriginal Palliative Care in British Columbia's Rural Interior." *Health & Social Care in the Community* 18 (5): 483–91. https://doi.org/10.1111/j.1365-2524.2010.00922.x.

Cherubini, Lorenzo. 2008. "The Metamorphosis of an Oral Tradition: Dissonance in the Digital Stories of Aboriginal Peoples in Canada." *Oral Tradition* 23 (2): 297–314.

Cranton, Patricia. 2002. "Teaching for Transformation." *New Directions for Adult and Continuing Education* 93 (1): 63–72. https://doi.org/10.1002/ace.50.

Cunsolo Willox, Ashlee, Sherilee L. Harper, and Victoria L. Edge, and the "My Word": Storytelling and Digital Media Lab, and Rigolet Inuit Community Government. 2013. "Storytelling in a Digital Age: Digital Storytelling as an Emerging Narrative Method for Preserving and Promoting Indigenous Oral Wisdom." *Qualitative Research* 13 (2): 127–47. https://doi.org/10.1177/1468794112446105.

Fletcher, Cristopher, and Carolina Cambre. 2009. "Digital Storytelling and Implicated Scholarship in the Classroom." *Journal of Canadian Studies. Revue d'Etudes Canadiennes* 43 (1): 109–30. https://doi.org/10.3138/jcs.43.1.109.

Freire, Paulo. 1971. *Pedagogy of the Oppressed.* New York: Continuum.

Glaser, Barney G. 1992. *Basics of Grounded Theory Analysis.* Mill Valley, CA: Sociology Press.

Godlewska, Anne, and Neil Smith, eds. 1994. *Geography and Empire.* Oxford: Blackwell.

Godlewska, Anne, Jackie Moore, and C. Drew Bednasek. 2010. "Cultivating Ignorance of Aboriginal Realities." *Canadian Geographer* 54 (4): 417–40. https://doi.org/10.1111/j.1541-0064.2009.00297.x.

Gubrium, Jaber F., and James A. Holstein, eds. 2003. *Postmodern Interviewing.* Thousand Oaks, CA: Sage. https://doi.org/10.4135/9781412985437.

Harris, R. Cole. 2002. *Making Native Space: Colonialism, Resistance, and Reserves in British Columbia.* Vancouver, BC: UBC Press.

Hart, Michael Anthony. 2010. "Indigenous Worldviews, Knowledge, and Research: The Development of an Indigenous Research Paradigm." *Journal of Indigenous Voices in Social Work* 1 (1): 1–16.

Heikkila, Karen Ann. 2007. "Teaching about Toponymy: Using Indigenous Place Names in Outdoor Science Camps." Master's thesis, University of Northern British Columbia, Prince George, British Columbia.

Hickling Hudson, Anne, and Peter Mayo. 2012. "Furthering the Discourse in Postcolonial Education." *Postcolonial Directions in Education* 1 (1): 1–8.

King, Thomas. 2003. *The Truth About Stories: A Native Narrative.* Toronto: House of Anansi.

Kollmuss, Anja, and Julian Agyeman. 2002. "Mind the Gap: Why do People Act Environmentally and What Are the Barriers to Pro-Environmental Behavior?" *Environmental Education Research* 8 (3): 239–60. https://doi.org/10.1080/13504620220145401.

Kreber, Carolin, and Patricia Cranton. 2000. "Exploring the Scholarship of Teaching." *Journal of Higher Education* 71 (4): 476–95. https://doi.org/10.1080/00221546.2000.11778846.

Lambert, Joe. 2008. *Digital Storytelling: Capturing Lives, Creating Community.* 2nd ed. Berkeley, CA: Center for Digital Storytelling.

Lambert, Joe, and the Center for Digital Storytelling. 2010. *Digital Storytelling Cookbook.* Berkeley, CA: Digital Diner Press.

Lavallée, Lynn F. 2009. "Practical Application of an Indigenous Research Framework and Two Qualitative Indigenous Research Methods: Sharing Circles and Anishnaabe Symbol-Based Reflections." *International Journal of Qualitative Methods* 8 (1): 21–40. https://doi.org/10.1177/160940690900800103.

Lawrence, Randee Lipson. 2008. "Powerful Feelings: Exploring the Affective Domain of Informal and Arts-Based Learning." *New Directions for Adult and Continuing Education* 120 (1): 65–77. https://doi.org/10.1002/ace.317.

Lincoln, Yvonna S., and Egon G. Guba. 1985. *Naturalistic Inquiry.* Beverly Hills, CA: Sage.

Louis, Renee Pualani. 2007. "Can You Hear Us Now? Voices from the Margins: Using Indigenous Methodologies in Geographic Research." *Geographical Research* 45 (2): 130–9. https://doi.org/10.1111/j.1745-5871.2007.00443.x.

McGregor, Deborah. 2004. "Coming Full Circle: Indigenous Knowledge, Environment and Our Future." *American Indian Quarterly* 28 (3): 385–410. https://doi.org/10.1353/aiq.2004.0101.

Mezirow, Jack. 2009. "Transformative Learning Theory." In *Transformative Learning in Practice: Insights from Community, Workplace, and Higher Education,* ed. Jack Mezirow, Edward W. Taylor, and Associates, 18–32. San Francisco: Jossey-Bass.

Mihesuah, Devan Abbott, and Angela Cavender Wilson, eds. 2004. *Indigenizing the Academy: Transforming Scholarship and Empowering Communities.* Lincoln, NE: University of Nebraska Press.

Moore, Janet. 2005. "Is Higher Education Ready for Transformative Learning? A Question Explored in the Study of Sustainability." *Journal of Transformative Learning* 3 (1): 76–91. https://doi.org/10.1177/1541344604270862.

Painter, Joe, and Alex Jeffrey. 2009. *Political Geography*. 2nd ed. Thousand Oaks, CA: Sage.

Parkes, Margot W. 2011. "Ecohealth and Aborignal Health: A Review of Common Ground." National Collaborating Centre for Aboriginal Health. Accessed 24 Feb. 2017. https://www.ccnsa-nccah.ca/docs/Ecohealth_Margot%20Parkes%202011%20-%20EN.pdf.

Pløger, John. 2001. "Public Participation and the Art of Governance." *Environment & Planning* 28 (2): 219–41. https://doi.org/10.1068/b2669.

Powell, Richard C. 2008. "Becoming a Geographical Scientist: Oral Histories of Arctic Fieldwork." *Transactions of the Institute of British Geographers* 33 (4): 548–65. https://doi.org/10.1111/j.1475-5661.2008.00314.x.

Ramsden, Paul. 1991. "A Performance Indicator of Teaching Quality in Higher Education: The Course Experience Questionnaire." *Studies in Higher Education* 16 (2): 129–50. https://doi.org/10.1080/03075079112331382944.

Said, Edward. 1978. *Orientalism*. London: Vintage Books.

Simpson, Leanne, ed. 2008. *Lighting the Eighth Fire: The Liberation, Resurgence, and Protection of Indigenous Nations*. Winnipeg: Arbeiter Ring.

Turner, Dale. 2006. *This Is Not a Peace Pipe: Towards a Critical Indigenous Philosophy*. Toronto: University of Toronto Press.

Van Eijck, Michiel, and Wolff-Michael Roth. 2007. "Keeping the Local Local: Recalibrating the Status of Science and Traditional Ecological Knowledge (TEK) in Education." *Science Education* 91 (6): 926–47. https://doi.org/10.1002/sce.20227.

Warry, Wayne. 2007. *Ending Denial: Understanding Aboriginal Issues*. Toronto: University of Toronto Press.

Whittemore, Robin, Susan K. Chase, and Carol Lynn Mandle. 2001. "Validity in Qualitative Research." *Qualitative Health Research* 11 (4): 522–37. https://doi.org/10.1177/104973201129119299.

Wright, Tarah. 2006. "Feeling Green: Linking Experiential Learning and University Environmental Education." *Higher Education Perspectives* 2 (1): 69–81.

DISCOVERING TRACES OF THE PAST ...

Palimpsest: "Layers of history." (Uncovered, embellished, reinvented, redefined. Sustainability appreciated through education, culture, and environments.)

Throughout history, Berlin has been a beautiful boil beneath Germany's thick skin. Having experienced tragedies of war such as poverty, destruction, segregation, and horrific antisemitism, Berlin still remains a resilient and ever-changing city. I found it awe inspiring to be in such an immensely developed, global capital that still seeks nature in every way possible. Even despite the fact that 13 million people currently inhabit this metropolis, there were tranquil urban gardens built from transformed airport runways and tiny urban courtyards where you could retreat from the chaos.

During our time in Berlin, I felt that this city taught me something about the true meaning of the "sustainability melting pot." Living in the heart of a global capital, it was incredible seeing so many bountiful community gardens, nature-farm schools, and city parks, emerging from demand, culture, tradition, and true love of nature.

Pentimento, as described by dictionary.com's "word of the day," is "the presence or emergence of earlier images, forms, or strokes that have been changed or painted over." I like this word because it builds on the strengths of "palimpsest" one step further by incorporating the concept of artistic expression over time.

In my opinion, the East Berlin wall is a symbol of sociocultural "sustainability" because of its ever-changing nature. It is an iconic piece of history, and yet the fact that it has been left unguarded from the public, open to graffiti and new artistic expressions, is admirable. I love the idea that a historically renowned mural of protest has been left to the public to decide its fate. It stands as a canvas for community to share their thoughts and expressions about the past with the world. It inspires living art for the future and reminds artists and audiences that while we can't change the past we have power in our pens to change the future.

Like "pentimento," the original murals on the East Side Gallery are beginning to fade with continued graffiti, tourists' signatures, and little notes over time, and yet the presence of earlier images still exists beneath the changes, reminding the present of its past.

This living mural touched my heart and made me think about the concept of sociocultural "sustainability" differently. It reminded me that people have the power to change a place and the impressions of its past, but they also have a prerogative to preserve it just the same. This mural reminded me that what I love most about a "sustainable" city is its ability to support freedoms of expression, share stories of change from the past, and inspire an even deeper sense of community into the future.

Sara Lax
Participant, University of Victoria's Northern
Europe Sustainability Field School

SECTION TWO

Implications of Place

The relationships among place, community, and learner/instructor in a field course are at the core of the learning experience. While it may seem obvious when thinking about field courses, the multifaceted importance of place and its meaning – both physical and social – is often unexplored by scholars reflecting on field school learning. Interactions with a specific place entail both visible and invisible significance. The pedagogical choices that instructors make may aim to uncover socio-economic, ecological, or political nuances in a particular community. At the same time, however, they may also serve to reinforce hierarchies of power and privilege. Although field course participants cannot ever "understand" a community (a concept that is, itself, clouded by one's own perspective), their interaction with place can reinforce or dispel stereotypes or introduce students to systems at work beyond the view of the casual visitor.

The chapters in Section 2 explore the importance of place in field school learning from different perspectives and in different contexts. John Borrows points to the centuries-old practice of place-based Indigenous legal education occurring outdoors, and draws parallels between student research methodologies that reconnect "outsider education" and learning through field trips and Indigenous law camps. He notes that learning law from and on the land can assist institutions in addressing the recommendations of Canada's Truth and Reconciliation Commission to provide training in intercultural competencies, conflict resolution, human rights, and anti-racism.

Deborah Curran's exploration of place focuses on the importance of field school learning for law students to gain perspectives about complex multiparty environmental disputes embedded in socio-ecological systems where Indigenous law continues to govern alongside colonial management regimes. Hearing from community members and those embedded in the place – government officials, First Nations staff, and Elders – enables students to come to appreciate the nuances of disputes and the possibility of resolution through multiple approaches rooted in different legal traditions. This is particularly important for understanding place-specific Aboriginal and Indigenous law and the continuing conflicts with federal and provincial colonial legal processes.

Finally, Elizabeth Vibert and Kirsten Sadeghi-Yekta explore the challenges facing instructors guiding students in community-based learning in the Global South. They reflect on instructor positionality, reminding readers that instructor perspectives and investments help to shape, in often unacknowledged ways, what students learn and how they interact with the community. Instructor choices about which community or people in a community to visit tend to privilege particular views; care is needed to ensure the visit does not simply confirm stereotypes (even positive ones). Vibert and Sadeghi-Yekta articulate the danger of short-term visits fostering a shallow view of community and point to the importance of giving back to the communities visited.

The authors of the chapters in Section 2 critically engage with the importance of place for learning, the care that must be taken in making choices about how students interact with place as a site of learning, and the significance of reciprocity when building relationships in the field.

The student vignettes in this section focus on the wonder of connecting with place and community through language and shared experience. Aisling Kennedy poignantly describes the importance of names for making intergenerational connections in an intercultural setting. Freya Selander reflects on the honour of living a life with meaning when faced with the shock of mass graves and other acts of xenophobia. Laura Buchan wonders at the importance of names and the kindness of strangers when trying to navigate transportation systems. Finally, Sarah Elwood describes the colour that lived experience in a community context injects into the bland, flat legal principles described in case law.

CONNECTING WITH THE COMMUNITY …

"Unga perru enya! Unga perru enya!"

This Tamil phrase was repeated to me as I sat with Elders Sundrabal and Jayamma in the courtyard of Tamaraikulam Elders Village. They repeated the words in loud strong voices as they shook their fingers and laughed. "Unga perru enya!" The Elders would say the phrase and then instruct me to repeat it myself. Even though I did not have a clue what they were asking me to say, I repeated the phrase until they were satisfied. Having met and spent much time with Sudrabal and Jayamma before, I connected with them that day to ask how to pronounce my students' names. Both of them patiently helped me pronounce beautiful Indian names such as Priyadarshini, Sowniya, Pushpalatha, Senthamarai, and Nisha. A kind woman walked by and witnessed my exchange with the Elders. She offered to translate for me. She told me that Pushpalatha means "flower," Santamarai means "lotus," and Nisha means "beauty." Growing up with a difficult name to pronounce myself, I knew how challenging it was when teachers would destroy and devalue my name in front of the class. I wanted the children to be proud of their name, their culture, and their history. I did not want the students to cater to my English tongue and change their names on my account. I would learn their names and show them they were important. When she finished translating each name, she paused and looked at me. "What is the meaning of your name?" I told her that my name was Gaelic for "beautiful dream." She then gave me a loving stare and asked, "What is your dream?" I was caught off guard – this seemed to be a very intimate question from a stranger, yet surprisingly I was relaxed and at ease. I took a moment and answered truthfully, "This is my dream. To do theatre around the world." She responded, "We are all very proud of you."

That night I reflected on my experience in the courtyard and could not escape the sound of the phrase, "Unga perru enya." It haunted me. I desperately wanted to know what it meant. After my next class at the Isha Foundation School, a few students stayed behind to chat. I took a chance and asked them if they could help me. I said the phrase again and asked them what it meant in English. They looked at me with bright eyes and answered, "What is your name?!" This was a delightful surprise. Through our intercultural and intergenerational exchange, the Elders were helping me communicate and ask the students their names in the classroom. In that moment it seemed everything was coming full circle. The Elders, the children, and I were connected through one simple yet powerful phrase, "Unga perru enya."

Aisling Kennedy
Participant, University of Victoria's
Applied Theatre Field School in India

5

Outsider Education: Indigenous Law and Land-Based Learning

JOHN BORROWS

Introduction

Legal education in North America once occurred outside the classroom (Llewellyn and Hoebel 1941). Before law schools were invented students apprenticed under the supervision of experienced practitioners.[1] They put in long hours. They learned through observation and practice (Hoebel 1974). Periodic lectures punctuated their experience, although in most jurisdictions they occurred infrequently. Examinations, on the other hand, were more common. Accrediting bodies ensured candidates understood the foundations of their field (Dewdney 1975). Credentials could be earned by degrees in some places (Berens and Hallowell 2009, 164). However, most education took place in proximity to senior members of the legal community.[2] It did not happen in lecture halls or seminar rooms. As such, it was more attentive to customary law (Berkes 1999; Angel 2002).

Of course, I am talking about Indigenous legal education in North America prior to European arrival. Indigenous law was taught on the land and water. Elders and other lawkeepers supervised their initiates in context. Demonstration, observation, and practice formed the heart of learning. Lectures could occur but emphasis was given to hands-on instruction. The form and substance of legal education was based on apprenticeships. Examinations were administered on a regular basis. Students were expected to recite songs, stories (cases), principles, teachings, and rules, in addition to demonstrating competence in ceremonial activities associated with the law. Degrees were awarded in societies; for instance, the Anishinaabe midewin society had a lifelong continuing education program (Hoffman 1891). Eight different degrees were awarded. They signified increasing power and expertise in Ojibwe law, as it was entwined with medicine, philosophy, and other areas of knowledge (Johnston 1982). Many of these systems still exist today (Benton-Banai 1988). They are a vibrant part of Canada's legal landscape (Borrows 2010; Akiwenzie-Damm 1996). Unfortunately, most law students, lawyers, legislators, and judges never learn of their existence. This must change.

Legal education should more fully engage land-based contextual learning (Black 2011). This would be wise regardless of the legal tradition being taught, such as the common law, civil law, or Indigenous law. Furthermore, more can be done in relation to outdoor learning when teaching the place of Indigenous law in Canada. Indigenous legal reasoning is often related to the land. In fact, understanding how Indigenous peoples practise law as a land-based activity is required to appreciate Canada's constitution. Indigenous law is an important source of authority for all Canadians in making decisions about our land and relationships. While some of this instruction can occur in the classroom, walls can hide important legal resources. Thus, law professors and students need to get outside more often. Again, this is true no matter the legal tradition. Indigenous legal education has something to offer North American law schools and the profession more generally. Participation and immersion in out-of-doors cultural contexts, even for short periods, can be very beneficial to learning. It also happens to be critical to understanding our broader place in this country.

This chapter discusses the importance of learning beyond the classroom in an Indigenous legal context. While university-based outdoor education is not unprecedented (Lowan 2008; 2009), it has been slow to develop in a law school environment (Wildcat et al. 2014). Law professors need to learn and apply this literature (Carlson, Lutz, and Schaepe 2009; Fuller, Edmondson, France, Higgitt, and Ratinen 2006; Menzies and Butler 2011; Castleden, Daley, Morgan, and Sylvestre, this volume; Curran, this volume), including its critiques (Nairn 2005; Hope 2009). Learning on and from the land should be integral to understanding our legal world. Law lives "on the ground"; it is present in physical and social relationships (Barsh 2008). Positivistic declarations of Parliament and the courts are not law's only source (Miller 2014). Legal obligations are generated in homes, businesses, hospitals, courts, cities, and rural landscapes (Macdonald 2002; Ellickson 1994, 2009; Sarat and Kearns 1995; Arthurs 1985). These and other legal sites should be explored and examined in more direct ways. Law professors can do more to mediate learning experiences in their immediate physical locations and much further afield.

In my view, land-based legal *site:ation* should take place in every law school across Canada. Law should be studied by directly experiencing and analysing law's interactions with the physical world. Scepticism and hard questions should form a vital part of this work (Pardy 2010). This would be fully consistent with the development of professional legal skills required of lawyers, judges, and other legal actors. Such experiences are also consistent with law schools' liberal arts heritage (Sarat 2004; Clark 2013–14). Learning in context should also help legal educators more fully embrace and critique insights from the sciences, humanities, and social sciences.[3] When you are on the land, it is more difficult to isolate legal phenomena from their broader context. Legal reasoning cannot be as easily contained within the four-square pages of case law, statutes, and other written texts.

At the same time, a call for outdoor legal education should not be seen as undermining classroom experience. This kind of learning should not replace other important methods employed in law schools. Balance is a watchword in Indigenous legal education. Holistic learning is encouraged in most Indigenous pedagogies. Lectures and seminars are valuable in their own right. They provide efficient ways to explore how cases are constructed. They help us probe how broader background forces inform the development of cases, legislation, and regulatory decisions. Moot courts, negotiations, and other simulations are also important tools in gaining practice-ready, hands-on experience. In addition, clinical legal education programs are essential in providing students with client-oriented education in a supervised setting. I have been in and around law schools for almost 30 years and participated in each mode of learning as a professor and student. I have found each of these methods to be very important in generating sound legal education.

At the same time, despite these important tools and pedagogies, more needs to be done. An imbalance exists in legal education. Law is not just about ideas; it is a practice. The life of the law is experience. Students are denied a well-rounded legal education if they spend their entire three years indoors. The often isolating walls of a law school can breed an unhealthy insularity. Law students can miss vital insights if professors pay insufficient attention to "cases" that lie beyond their doors. Furthermore, student interest and engagement can be significantly heightened by field school moments.

In examining these issues, this chapter first discusses how law is currently being taught outdoors in a few notable examples. Second, it examines how Indigenous legal methodologies contain somewhat unique insights for learning law on and from the land. Third, I discuss my own experience in teaching Anishinaabe law in an outdoors context to demonstrate how students can develop deeper understandings of their professional responsibilities. In this regard, I consider the group phenomena of Indigenous Law Camps and the individual experiences of guided study in more isolated settings. Throughout this chapter the central place of telling stories in outdoor locations as a means of recording and transmitting law is never far from the surface.

Learning from the Land: Understanding Indigenous Pedagogies

Outdoor legal education also provides significant room for developing pedagogies that are attentive to Indigenous legal traditions.[4] It is critical to remember that these pedagogies will not be appropriate to all law school courses. Nevertheless, they can be an important part of a school's repertoire for increasing student insight and skills. Like a few grains of yeast in the preparation of bread, the addition of Indigenous law's methodologies may positively lift the entire institution. This chapter does not suggest that Indigenous law "take over" any law school's legal curriculum. I am only trying to put Indigenous law in its proper "place." While the intensity of engagement will vary, all law schools should consider ways in which they can teach Indigenous law, which after all, is a vital part of the laws of Canada (Borrows 2002).

Thus, in enhancing Indigenous education, each law school would do well to pay attention to the Indigenous Nations around them, even in urban settings. Students and faculty will find significantly greater opportunities for engagement if they respectfully develop courses and programming with practitioners, communities, and teachers who are Indigenous to the school's immediate physical setting. Learning about local Indigenous law will help students interact differently with their surrounding physical environments throughout their three-year enrolment. Even when a law school has a national student body, or focuses on Indigenous law more generally, it is equally vital that local Indigenous legal insights be taught, studied, applied, critiqued, and experienced in direct as well as more abstract ways. This will reinforce the notion that law does not just flow from legislatures, courts, administrative bodies, or other positivistic processes. Law is also sourced in specific First Nations, Metis, and Inuit legal systems, which give rise to obligations and rights in particular contexts.

Distinctive Indigenous legal methodologies can be further developed and applied to assist students in learning law on and from the land (Cajete 1994). There are many examples regarding Indigenous law being sourced in the land (Black 2011). When I speak of land, I am also speaking of all the natural phenomena associated with the land, such as water (Bedard 2010).

One prominent declaration of land-based laws was evident during the Gitksan and Wet'suwet'en Hereditary Chiefs Opening Statement in the Delgamuukw case. As readers will recall, this case tested the existence of Aboriginal title in British Columbia (*Delgamuukw* v. *British Columbia* 1997). In outlining how law is sourced in the land, Chiefs Gisday Wa and Delgam Uukw said:

> For us, the ownership of territory is a marriage of the Chief and the land. Each Chief has an ancestor who encountered and acknowledged the life of the land. From such encounters comes power. The land, and plants, the animals and the people all have spirit, they all must be shown respect. That is the basis of our law.
>
> The Chief is responsible for ensuring that all the people in his House respect the spirit in the land and in all living beings ... My power is carried in my House's histories, songs, dances and crests. It is recreated at the Feast when the histories are told, the songs and dances performed and the crests displayed. With the wealth that comes from the respectful use of the territory, the House feeds the name of the Chief in the feast hall. In this way, the law, the Chief, the territory, and the Feast become one ... By following the law, the power flows from the land to the people through the Chief, by using the wealth of its territory. ("The Address of the Chiefs, May 11, 1987," 1992, 22)

This statement demonstrates how Gitksan and Wet'suwet'en law flows from their relationship with the earth. The Delgamuukw case was built from this perspective (Mills 1994; 2005). Outside the case, the Chiefs use the land as a legal resource for strengthening their House and its relationships (Daly 2005). The histories, songs, crests, and food from the land are organized in particular ways (Mills 1994; 2005). Totem poles are erected to encode these histories and crests, and they are purposely rooted in the earth to communicate a House's law (Wa and Uukw 1989, 26). They are marshalled as authority for the House's past and future actions (Overstall 2005).

Reading the earth, through poles, songs, clans, and histories is not unique to the Gitxsan and Wet'suwet'en. Similar appeals to the earth, although culturally distinct to each group, can be found in most Indigenous legal traditions (Austin 2009; Richland 2008). The Cree talk about *miyo-wicehtowin* and *witaskewin* (Cardinal and Hildebrandt 2000, 14–20, 39–42). The Blackfoot discuss the concepts of *kakyosin* and *mokaksin* (Bastien 2004, 119–27). The Haudenosaunee relate their Great Law of Peace to pine trees, eagles, and other natural phenomena (Fenton 1998). The Mi'kmaq source Netukulimk (Prosper et al. 2011; Henderson 1995a, 1995b; Sable and Francis 2012). The Secwepemc value the natural-world sourcing of their laws: *yiri7 re stsq'ey's kucw* (Ignace 2008). The Hul'q'umi'num'[5] advance the idea of *snuw'uyulth* (Paige 2004, 1). The Haida apply *Yah'gudaan* (Quail 2014).

My own community, the Chippewas of the Nawash at Neyaashiinigmiing, also regards the earth as an important legal resource. The Anishinaabe have a tradition of practising law by reference to the natural world (Greenwood and de Leeuw 2007). The Anishinaabe word for this concept is *gikinawaabiwin* (Brenda Fairbanks, pers. comm., 7 Jan. 2014). The word *akinoomaage* is also used to describe this phenomenon. It is formed from two roots: *aki* and *noomaage*. "Aki" means earth and "noomaage" means to point towards and take direction from. The idea this word conveys is that analogies can be drawn from our natural surroundings and applied to or distinguished from human activity. This is the heart of Anishinaabe legal reasoning: parallel situations are correlated, dissimilar situations are distinguished. In this legal approach, the environment becomes the legal archive that practitioners read and use to regulate their communities.[6]

In this legal tradition, the earth has a culture that the Anishinaabe strive to embed in their laws. These principles may often be found in the Anishinaabe language. For instance, there is a time in the early Ontario spring when cold and warm air masses intermingle, causing fine mists to rise over the earth. The word used to describe this phenomenon is *aabawaa*, which means warm and mild. At these moments winter starts to loosen her grip on the land. The snows melt and waters start to flow. Sap can begin running through the trees as nature prepares to nurture new life. Interestingly, the Anishinaabe word for forgiveness – *aabawaawendam* – is related to this moment in time. Thus, forgiveness can be analogized to loosening one's thoughts towards others, to letting relationships flow more easily, with fewer restrictions. Forgiveness is a state of being warmer and milder towards another; it signifies a warming trend in a relationship. Notice that forgiveness, like the clearing of early spring mists, does not occur in an instant. Heat and the warmth need to be applied through a sustained period of time for mists to clear. Clarity of vision takes a while to develop as spring mists do not dissipate immediately; time is often needed to "clear the air" and bring fairer views. So it is with forgiveness.

This example provides a glimpse into understanding how reasoning about and practising law can be related to "reading the land." Chief Gary Potts of the Temagami Anishinaabe exemplified Anishinaabe legal reasoning related to the land through an experience he shared from his territory (Borrows 2015). He wrote:

> I remember once coming across an old white pine that had fallen in the forest. In its decayed roots a young birch and a young black spruce were growing, healthy and strong. The pine tree was returning to the earth, and two totally different species were growing out of the common earth that was forming. And none was offended in the least by the presence of the others because their own identities were intact.
>
> When you walk in a forest you see many forms of life, all living together. They each have their own integrity and the capability to be different and proud. I believe there is a future for native and non-native people to work together because of the fundamental fact that we share the same future with the land we live on.
>
> We will never be able ... to build another planet like earth or build a covered bridge to another planet and start all over again. We need to acknowledge that the *land is the boss*. (Potts 1992, 199; original emphasis)

Chief Potts's attempt to describe what he saw in the bush and draw meaning from his experience is an example of *gikiniwaabiwin* and *akinoomaagewin*. This is an example of physical philosophy which is aimed at drawing law from real-world experiences.

In addition to direct observation of the land, stories related to the land also play an important role in transmitting legal knowledge.[7] Basil Johnston (2002, 17–50; 1982, 165–6; 1976, 151–3), an Elder from my community who passed away in 2015, often talked about how Anishinaabe law is learned by studying the earth. He often told stories about insects, birds, and animals to illustrate the rights and obligations of our Band Council, Chiefs, and general community members (Johnston 1995). He tried to impress on me that, as a law teacher, I needed to study the earth and learn her stories to find principles which could be applied to more

effectively regulate our behaviour (Johnston 2012). As I searched his
writings, I came across a quote in a related vein wherein he wrote:

> Learning comes not only from books but from the earth and our sur-
> roundings as well. Indeed, learning from the mountains, valleys, forest and
> meadows anteceded book knowledge. What our people know about life
> and living, good and evil, laws and the purposes of insects, birds, animals
> and fish comes from the earth, the weather, the seasons, the plants and
> the other beings. The earth is our book; the days its pages; the seasons,
> paragraphs; the years, chapters. The earth is a book, alive with events that
> occur over and over for our benefit. Mother earth has formed our beliefs,
> attitudes, insights, outlooks, values and institutions. (Johnston 2003, v)

Learning from the earth in this way takes time. The idea that laws (among
other human activities) can be learned from reading the earth requires
practice. It requires a literacy that is not often taught and developed
in a university setting.[8] Developing such land-based literacy should be
adopted as a more explicit goal within Canadian law schools.

My Experience: Indigenous Legal Education

One way in which students are being slowly introduced to Indigenous
land-based laws is through spending short periods of time on the land
with Indigenous legal practitioners. There is a sustained phenomenon
developing in a few law schools that places law students in Indigenous
communities for intensive four-day experiences. In addition, Indigenous
students are basing their graduate work on legal relationships emanating
from the land in their home communities.

AbCamp – Coast Salish and
Kwakwaka'wakw Legal Traditions

The University of Victoria Faculty of Law has been doing this camp for
twenty years. Called AbCamp, it was initiated in 1995 by student Ann
Roberts, who was an undergraduate student at the University of Calgary
when she heard of a similar camp held by the Peigan and Siksika people
of the Blackfoot Nation for the RCMP.[9] The camp is organized by the students with the support of a faculty director and takes place each year in
conjunction with local First Nations. During their time in these communities, students participate in ocean canoe voyages, pit cooking, singing,

drumming, storytelling, games of lahal, cedar-bark weaving, hiking, and other traditional activities. During its first few years, the AbCamp experience was an opportunity to break down barriers between law students and local communities. In recent years, now that relationships are more fully formed, ideas and practices related to Indigenous law are implicitly and explicitly woven into the four-day gathering. Alumni continually report that it is one of the best experiences they have during law school. For instance, lawyer Claire Truesdale wrote:

> As a UVic alum, I attended the program myself. It is an introduction for new UVic law students to the cultures of the peoples in whose traditional territories they will be living and studying colonial law. But more than that, it is an eye opening and unsettling opportunity to learn about the modern reality of Vancouver Island First Nations. It is, perhaps most importantly, the barest of introductions to the legal traditions of the Island's indigenous peoples. It is a rare occasion for indigenous and non-indigenous peoples to come together in an atmosphere of respect and willingness to learn. The elders who addressed the students and guests on Saturday night spoke of the need to be able to look each other in the eye, talk to each other and respect one another. Only then, they said, can we begin to move towards healing the rifts between our communities. (Pers. comm.) [10]

In a similar vein, UVic alum and lawyer Berry Hykin wrote:

> AbCamp gives you a rare opportunity at the beginning of your legal education to have a hands-on experience and obtain a different perspective on law and its practical aspect ... It challenges you to see how the law actually affects people in real and profound ways. You carry that perspective with you all the way through law school so it can inform how you approach your studies and how you eventually approach your practice. The impact of First Nations, historically and currently, on our society is so profound. Especially for people going into the legal profession, I think it's extremely important that this vital part of our community is made tangible through the participatory experience of something like AbCamp. [11]

In recent years, students from the University of British Columbia were invited to attend UVic's AbCamp to allow them to enjoy the experience between schools and work on ways to develop their own program. As a result, UBC has just completed its own second annual Indigenous Law Camp. [12] Likewise, Osgoode Hall Law School has also completed its second annual Anishinaabe Law Camp, and I will now address that experience.

Anishinaabe Law Camp: Working with Groups

The Anishinaabe Law Camp with Osgoode Hall Law School is hosted by the Chippewas of the Nawash First Nation at Neyaashiinigmiing on the shores of the Saugeen Peninsula in Ontario, which is my home community. My family and I play a large role in its development and implementation. The program was initiated by Professor Andrée Boiselle, who had been a student at UVic Law School. She had participated in UVic's AbCamp while working towards her PhD in Victoria. She had also participated with UVic Professor John Lutz and University of Saskatchewan Professor Keith Carlson in their Sto:lo Field School while at UVic (Carlson, Lutz, and Schaepe 2009). This field school took place in the Fraser Valley and was hosted by Dr Sonny McHalsie and others within the Sto:lo community. When Andrée started teaching at Osgoode Hall Law School she found herself searching for ways to develop similar opportunities for students in central Canada. Thus, Andrée approached me and my family with a request that we consider ways in which we might introduce Osgoode Hall law students to Anishinaabe law.

We identified that the purpose of the camp is to facilitate student understanding of Anishinaabe law, by teaching and learning about how to regulate our behaviour and resolve disputes. The curriculum is organized to do this through introducing students to different sources of law at Neyaashiinigmiing. The formal curriculum had the students participate in activities related to air, fire, water, earth, plants, animals, and fish. Each activity took place at a different venue to allow students to learn about the laws that flowed from each of these entities. Stories, songs, lectures, demonstrations, and hand-on activities were used in various ways. We gathered by lakeshores, huddled together around fires, walked through dense hardwood forest, paddled across waters, and sprawled across our powwow grounds and in our community centres. Students experienced pipe ceremonies, prepared medicine bundles, case-briefed Anishinaabe stories, and handled treaty wampum. They were the beneficiaries of unplanned appearances by local community members. It is this immersion experience in the context of a particular ecology and community that cannot be replicated in the classroom.

Each year approximately fifty Osgoode Hall law students and professors attend the camp. A number of alumni, visiting professors, and friends also attend the events. Furthermore, community support is also strong as people pitch in, drop in to listen, or offer behind-the-scenes support. Before the first camp, Elder Isabel Millette was given tobacco in accordance with our legal processes in connection with selecting a name for the camp. Through study and further offering she also received a motto for the camp, which is *Pii dash Shkakimi-kwe giigidid aabdeg gbizindawaamin* – When the earth speaks we will listen.

Law Faculty participants positively shared their thoughts concerning the program:

[Andrée Boiselle] appropriately articulated the goals of the Camp to include not just exposing the participants to Indigenous intellectual resources, pedagogies, and modes of reasoning, but also "laying the ground," experientially and relationally, for the participants' further engagement with Indigenous political and legal traditions. In Andrée's analysis, this is like "planting a seed," so that the activities undertaken at the Camp would inform the participants' approach to cross-cultural legal conversations and to the ongoing work of decolonization. She envisioned it as giving shape to their responsibility as legal scholars and practitioners to learn about and take account of Indigenous laws and perspectives.

In practice, once on the land the participants in the Camp embraced these teachings. The program was carefully laid out so as to give students the opportunity to hear stories rooted in particular places, experiences and relations, and these stories drew their strength from the land and water in whose company they were shared. The impact of the stories on the students was visibly obvious, and was reflected in the sharing circle held at the end of the Camp. Many times, we heard someone refer to a particular story by referencing where the storyteller was at the time the story was told, and this is further evidence of the critical importance and pedagogical value of doing this type of experiential teaching on the land.

The benefits to students and faculty alike who attended the Camp cannot be easily summarized. As witnessed at the sharing circle, the diversity of lessons, teachings, and emotions experienced by participants was broad and deep. People shared jokes, songs, tears, and stories with one another in a heartfelt and genuine manner. Students and faculty spoke of the people they had met and learned from, such as the firekeepers, Elders, and Borrows family teachers, including John and Lindsay. They spoke of the lessons they would take back, both personally and professionally, as well as a commitment to continue the work that had been started and to share their experiences with other students and their families. Looking at the limestone escarpments, seeing the replica wampum belts, and hearing of precolonial and post-Confederation treaties left an indelible impression of the maturity of the relationships that we had all stepped into in different ways as citizens, immigrants, and visitors to these lands and waters. It is fair to say that this impression was met with a sense of responsibility, too, by student and faculty participants alike. While there has been a recent flourishing of scholarship around Indigenous laws and legal traditions, it's clear that much work remains to be done for our textbooks, course materials, students, and faculty to reflect the knowledge and importance of Indigenous laws as sources of authority relevant to everyone. (Bhatia, Buchanan, Imai, McNeil, Shanks, Scott, and Hennessy, pers. comm.)

Guided Sole Practitioner Experiences: Working with Individuals

In addition to facilitating group-based legal educational experiences, there is also room for taking individual students into communities and learning law from experienced Indigenous practitioners. I have seen this occur most strongly in the context of graduate students, although J.D. students have also benefited from these interactions on a more ad hoc basis.

For example, Professor Sarah Morales worked closely with Hul'q'umi'num' lawkeepers on Vancouver Island during the half-dozen years of her graduate work. By following Hul'q'umi'num' legal process learned under the guidance of experienced practitioners, Sarah discovered how legal principles were embedded in the land and reinforced with stories. In describing this process, she wrote:

Through the land you can know who my community is and the shape of our legal traditions. Land is the very entity that has inspired, recorded and preserved our histories ... Through my educational journey, I have learned that "places possess a marked capacity for triggering acts of self-reflection, inspiring thoughts about who one presently is, or memories of who one used to be, or musings on who one might become." Different landscapes conjure different emotions. My territory prompts reflection upon what was "experienced" or "learned" in those places ... As I spent my days poring over books, learning about the history of my community and listening to stories about Our First Ancestors, I found my source of strength – the land. I began to picture my traditional territory in a different way ... Our laws were written in the lands – quietly and majestically surrounding us, waiting for their renewal. These are the laws that can bring healing and health to my people. As a result, when I now return home to my community I don't focus on the over-development of our territory and the poverty of our reserve lands. Instead I look up towards Swuq'us; I see the face of my First Ancestor Stutson and remember the teachings he gave me. This realization is what I hope to share with my community through this dissertation. I want to change how they envision our landscape. I want them to draw strength from the laws embedded in the lands. I want them to recognize that we are a strong people. I want them to see we have a living legal tradition that operates within our communities and guides our daily decision-making processes. (Morales 2014, 36, 41, 42)

Danika Littlechild also had a similar experience of learning from the land through drawing on Cree legal traditions during her LL.M. experience. Throughout her thesis, Danika describes the laws she learned in regard to water, and she uses this information to work towards reconciliation with municipal, provincial, and federal governments. In taking this course, she regards Cree law as a vital element of this reconciliation. In this light, Danika's thesis is rooted in Cree "laws [which] have developed as a result of observations of the natural environment, and over centuries of interactions between *ayisiyinowak* people / human beings and *okâwîmâwaskiy* mother earth" (Littlechild 2014, 27). Thus, she writes:

Nipiy is the Cree word for water. "Ni" derives from niya, meaning "I" or "I am." "piy" derives from the word pimatisiwin, meaning "Life." Nipiy is thus properly understood as meaning "I am Life." Water is lifeblood, animating us as human beings, and all that is around us. The Cree language operates on the principle of anima, life-force. Understanding that elements of ourselves and our environment(s) have an inner life force determines how those elements are described, usually in a relational manner. Water is as much a process as it is an entity. Water has so many identities in our language – over 40 words or phrases in Cree describe water in all its forms and manifestations. Water is a living, cultural and spiritual entity that defies reduction to a mere resource.

   The presence of spirit in water and its place in our lives is tied to the way Cree Legal Traditions describe rights and duties affiliated with water. (Littlechild 2014, 19, 20, 24)

The thesis demonstrates how learning law from the land and water within a specific First Nation tradition can generate creative solutions to meet our contemporary legal needs.

Conclusion

Experiences learning law from and on the land are recorded in constitutions, statutes, regulations, bylaws, declarations, adjudicative judgments, songs, carvings, textiles, dances, wampum belts, scrolls, petroglyphs, etc. Stories are also an important vehicle for recording such laws (Green 2014). In her LL.M. thesis, Secwepemc lawyer Nancy Sandy (2011, 89) put it this way: "Law is also embedded in stories. Like common law cases, they could communicate appropriate and inappropriate behavior. Stories could also record punishments, or chronicle when mercy or justice was extended or retracted." Hadley Friedland (2012), who worked with Cree stories in both her LL.M. and PhD theses, similarly observed that these stories could help students attend to law "on the ground" within Indigenous communities.

The Indigenous Law Research Unit (ILRU) at the University of Victoria focused on stories as a resource for legal reasoning and practice. The ILRU has worked with band councils, Elders, hereditary chiefs, clan mothers, and other customary structures to understand these legal traditions. Law students, lawyers, judges, and legal academics have worked with these individuals and groups to "honour the internal strengths and resiliencies present in Indigenous societies and in their legal traditions, and to identify legal principles that may be accessed and applied today – to governance, lands and waters, environment and resources, justice and safety, and building Indigenous economies."[13] Professor Val Napoleon, who is the director of this unit, has done groundbreaking work in bringing Indigenous law more fully to light. Legal education that attends to other dimensions of Indigenous law is being developed by other law professors across the country and their varied approaches are vital to this work.

The information contained in this chapter shows what law schools, legal educators, and governments might do in fulfilling the recommendations of the Indian Residential Schools Truth and Reconciliation Commission. Recommendations 27 and 28 call for law schools to create a course in the laws related to Aboriginal peoples, skills-based training in intercultural competency, conflict resolution, human rights, and anti-racism (2015, 168). Recommendation 50, under "Equity for

Aboriginal People in the Legal System," calls on the federal government "to fund the establishment of Indigenous law institutes for the development, use, and understanding of Indigenous laws and access to justice" (207).

NOTES

This chapter was originally published as John Borrows (2016), "Outsider Education: Indigenous Law and Land-Based Learning," *Windsor Yearbook of Access to Justice* 33 (1): 1–27, and is reprinted here with permission in a slightly shortened version that includes minor editorial changes. I would like to thank the following individuals for their helpful suggestions on earlier drafts: Deborah Curran, Karen Drake, Douglas Harris, John Kleefeld, Damien Lee, and Aaron Mills.

1 The title for this task in Anishinaabemowin is "oshkaabewis." For a description of what is learned in this role, see Mary Siissip Geniusz (2015). Mary was a traditionally trained apprentice of the late Keewaydinoquay (see Keewaydinoquay 2006).
2 For examples, see Cruikshank (2005), Wright (1962), and Fiske and Patrick (2000).
3 A prominent call for interdisciplinary legal education was found in *Law and Learning by Consultative Group on Research and Education in Law* (Social Sciences and Humanities Research Council of Canada 1983) [The Arthurs Report]. For an evaluation, see Backhouse (2003, 33) and Arthurs and Bunting (2014, 487).
4 For a deep discussion about land-based pedagogy in one Indigenous context, see Basso (1996).
5 There are two commonly accepted spellings (Hul'qumi'num and Hul'q'umi'num') and both are used in this book.
6 For a discussion of how rocks can be read and marked, see Morton and Gawboy (2000) and Rajnovich (1994). For a discussion of how animals can be read, see Bohaker (2010) and Pomedli (2014).
7 For further development of this point, see Noodin (2014).
8 For an explanation of how Indigenous literacy might be better developed, see Michell (2014).
9 For more information, see UVic Law's "AbCamp: 20 Years of Understanding" at https://www.uvic.ca/law/home/news/archive/AbCamp20thAnniversary .php – Aboriginal Awareness Camp 2011. 28 Jan. 2012. https://www.youtube .com/watch?v=P6B8ViCDYLI.

10 See "UVic Law's AbCamp Celebrates Its 20th Anniversary." 6 Oct. 2015. http://jfklaw.ca/uvic-laws-abcamp-celebrates-its-20th-anniversary/.

11 See UVic Law's "AbCamp: 20 Years of Understanding" at https://www.uvic .ca/law/home/news/archive/AbCamp20thAnniversary.php.

12 For more information, see UBC Peter Allard School of Law's Indigenous Awareness Camp at http://www.allard.ubc.ca/indigenous-legal-studies -program/indigenous-student-support/indigenous-legal-studies-program -events.

13 For more information, see UVic Law's Indigenous Law Research Unit at https://www.uvic.ca/law/about/indigenous/indigenouslawresearchunit.

REFERENCES

"[The] Address of the Chiefs, May 11, 1987." 1992. In *Colonialism on Trial: Indigenous Law Rights and the Gitksan and Wet'suwet'en Sovereignty* Case, edited by Don Monet and Skanu'u (Ardthye Wilson), 22–3. Philadelphia, PA, and Gabriola Island, BC: New Society.

Akiwenzie-Damm, Kateri. 1996. "We Belong to This Land: A View of 'Cultural Difference.'" *Journal of Canadian Studies. Revue d'Etudes Canadiennes* 31 (3): 21–8. https://doi.org/10.3138/jcs.31.3.21.

Angel, Michael. 2002. *Preserving the Sacred: Historical Perspectives on the Ojibway Midewiwin*. Winnipeg: University of Manitoba Press.

Arthur's Report. See Social Sciences and Humanities Research Council of Canada.

Arthurs, Harry W. 1985. *"Without the Law": Administrative Justice and Legal Pluralism in Nineteenth-Century England*. Toronto: University of Toronto Press.

Arthurs, Harry W., and Annie Bunting. 2014. "Socio-Legal Scholarship in Canada: A Review of the Field." *Journal of Law and Society* 41 (4): 487–99. https://doi.org/10.1111/j.1467-6478.2014.00682.x.

Austin, Raymond D. 2009. *Navajo Courts and Navajo Common Law: A Tradition of Tribal Self-Governance*. Minneapolis: University of Minnesota Press.

Backhouse, Constance. 2003. "Revisiting the Arthurs Report Twenty Years Later." *Canadian Journal of Law and Society* 18 (1): 33–44. https://doi.org/10.1017/S0829320100007432.

Barsh, Russel L. 2008. "Coast Salish Property Law: An Alternative Paradigm for Environmental Relationships." *Hastings West-Northwest Journal of Environmental Law and Policy* 14: 1375–416.

Basso, Keith. 1996. *Wisdom Sits in Places: Landscape and Language among the Western Apache*. Albuquerque: University of New Mexico Press.

Bastien, Betty. 2004. *Blackfoot Ways of Knowing*. Calgary: University of Calgary Press.

Bedard, Renée Elizabeth Mzinegiizhigo-kwe. 2010. "Keepers of the Water: Nishnaabe-kwewag Speaking Out for the Water." In *Lighting the Eighth Fire: The Liberation, Resurgence, and Protection of Indigenous Nations*, ed. Leanne Simpson, 89–110. Winnipeg: Arbeiter Ring.

Benton-Banai, Edward. 1988. *The Mishomis Book: The Voice of the Ojibway*. St Paul: Red School House Publishers.

Berens, William, as told to A. Irving Hallowell. 2009. *Memories, Myths, and Dreams of an Ojibwe Leader*, edited by Jennifer S.H. Brown and Elaine Gray. Montreal: McGill-Queen's University Press.

Berkes, Fikret. 1999. *Sacred Ecology: Traditional Ecological Knowledge and Resource Management*. Philadelphia: Taylor & Francis.

Black, Christine F. 2011. *The Land Is the Source of the Law: A Dialogic Encounter with Indigenous Jurisprudence*. New York: Routledge.

Bohaker, Heidi. 2010. "Reading Anishinaabe Identities: Meaning and Metaphor in *Nindoodem* Pictographs." *Ethnohistory (Columbus, Ohio)* 57 (1): 11–33. https://doi.org/10.1215/00141801-2009-051.

Boisselle, Andrée. 2010. "Beyond Consent and Disagreement: Why Law's Authority Is Not Just about Will." In *Between Consenting Peoples: Political Community and the Meaning of Consent*, ed. Jeremy H.A. Webber and Colin Murray Macleod, 207–32. Vancouver: UBC Press.

Borrows, John. 2002. "Fourword: Issues, Individuals, Institutions, and Ideas." *Indigenous Law Journal at the University of Toronto Faculty of Law* 1 (1): vii–xviii.

Borrows, John. 2010. *Canada's Indigenous Constitution*. Toronto: University of Toronto Press.

Borrows, John. 2015. *Freedom and Indigenous Constitutionalism*. Toronto: University of Toronto Press.

Cajete, Gregory. 1994. *Look to the Mountain: An Ecology of Indigenous Education*. Durango: Kivaki Press.

Cardinal, Harold, and Walter Hildebrandt. 2000. *Treaty Elders of Saskatchewan*. Calgary: University of Calgary Press.

Carlson, Keith, John Lutz, and David Schaepe. 2009. "Turning the Page: Ethnohistory from a New Generation." *University of the Fraser Valley Research Review* 2 (2): 1–8.

Clark, Sherman J. 2013–2014. "Law School as Liberal Education." *Journal of Legal Education* 63: 235–46.

Cruikshank, Julie. 2005. *Do Glaciers Listen? Local Knowledge, Colonial Encounters, and Social Imagination*. Vancouver: UBC Press.

Daly, Richard. 2005. *Our Box Was Full: An Ethnography for the Delgamuukw Plaintiffs*. Vancouver: UBC Press.

Delgamuukw v. British Columbia, 3 S.C.R. 1010 (1997).

Dewdney, Selwyn. 1975. *The Sacred Scrolls of the Southern Ojibway*. Toronto: University of Toronto Press.

Ellickson, Robert. 1994. *Order without Law: How Neighbors Settle Disputes.* Cambridge, MA: Harvard University Press.

Ellickson, Robert. 2009. *The Household: Informal Order Around the Hearth.* Princeton: Princeton University Press.

Fenton, William N. 1998. *The Great Law and the Longhouse: A Political History of the Iroquois Confederacy.* Norman: University of Oklahoma Press.

Fiske, Jo-Anne, and Betty Patrick. 2000. *Cis Dideen Kat (When the Plumes Rise): The Way of the Lake Babine Nation.* Vancouver: UBC Press.

Friedland, Hadley. 2012. "Reflective Frameworks: Methods for Accessing, Understanding, and Applying Indigenous Laws." *Indigenous Law Journal at the University of Toronto Faculty of Law* 11 (1): 1–40.

Fuller, Ian, Sally Edmondson, Derek France, David Higgitt, and Ilkka Ratinen. 2006. "International Perspectives on the Effectiveness of Geography Fieldwork for Learning." *Journal of Geography in Higher Education* 30 (1): 89–101. https://doi.org/10.1080/03098260500499667.

Geniusz, Mary Siisip. 2015. *Plants Have so Much to Give Us, All We Have to Do Is Ask: Anishinaabe Botanical Teachings.* Minneapolis: University of Minnesota Press.

Green, Jacquie Kundoqk. 2014. "Transforming Our Nuuyum: Contemporary Indigenous Leadership and Governance: Stories Told by Glasttowk askq and Bakk jus moojillth, Ray and Mary Green." *Indigenous Law Journal at the University of Toronto Faculty of Law* 12 (1): 33–60.

Greenwood, Margo, and Sarah de Leeuw. 2007. "Teachings from the Land: Indigenous People, Our Health, Our Land, and Our Children." *Canadian Journal of Native Education* 30 (1): 48–53.

Henderson, James (Sakej) Youngblood. 1995a. "First Nations Legal Inheritances: The Mikmaq Model." *Manitoba Law Journal* 23: 1–31.

Henderson, James (Sakej) Youngblood. 1995b. "Mikmaw Tenure in Atlantic Canada." *Dalhousie Law Journal* 18: 196–294.

Hoebel, E. Adamson. 1974. *The Law of Primitive Man.* New York: Atheneum.

Hoffman, Walter James. 1891. "The Midewiwin, or "Grand Medicine Society," of the Ojibwa in Smithsonian Institution." *U.S. Bureau of Ethnology Report* 7: 149–299.

Hope, Max. 2009. "The Importance of Direct Experience: A Philosophical Defence of Fieldwork in Human Geography." *Journal of Geography in Higher Education* 33 (2): 169–82. https://doi.org/10.1080/03098260802276698.

Ignace, Ronald Eric. 2008. "Our Oral Histories Are Our Iron Posts: Secwepemc Stories and Historical Cconsciousness." PhD diss., Simon Fraser University, Vancouver, British Columbia.

Johnston, Basil. 1976. *Ojibway Heritage.* Toronto: McClelland and Stewart.

Johnston, Basil. 1982. *Ojibway Ceremonies.* Toronto: McClelland and Stewart.

Johnston, Basil. 1995. *The Bear Walker and Other Stories.* Toronto: Royal Ontario Museum.

Johnston, Basil. 2002. *The Manitous: The Spiritual World of the Ojibway.* Toronto: Key Porter Books.

Johnston, Basil. 2003. *Honour Earth Mother: Mino-Adujaudauh Mizzu-Kumik-Quae.* Cape Croker, ON: Kegedonce Press.

Johnston, Basil. 2012. *Living in Harmony: Mino-nawae-indawaewin.* Cape Croker, ON: Kegedonce Press.

Keewaydinoquay, Peschel. 2006. *Stories from My Youth,* edited by Lee Boisvert. Lansing: Michigan State University Press.

Littlechild, Danika Billie. 2014. *Transformation and Re-formation: First Nations and Water in Canada.* Victoria, BC: University of Victoria Press.

Llewellyn, Karl N., and E. Adamson Hoebel. 1941. *The Cheyenne Way: Conflict and Case Law in Primitive Jurisprudence.* Norman: University of Oklahoma Press.

Lowan, Greg. 2008. "Outward Bound Giwaykiwin: Wilderness Based Indigenous Education." M.A. thesis, Lakehead University, Thunder Bay, Ontario.

Lowan, Greg. 2009. "Exploring Place from an Aboriginal Perspective: Considerations for Outdoor and Environmental Education." *Canadian Journal of Environmental Education* 14: 42–58.

Macdonald, Roderick A. 2002. *Lessons of Everyday Law.* Montreal: McGill-Queen's University Press.

Menzies, Charles R., and Caroline F. Butler. 2011. "Collaborative Service Learning and Anthropology with Gitxaała Nation." *Collaborative Anthropologies* 4 (1): 169–242. https://doi.org/10.1353/cla.2011.0014.

Michell, Herman J. 2014. *Working with Elders and Indigenous Knowledge Systems: A Reader and Guide for Places of Higher Learning.* Vernon, BC: Charlton Publishing.

Miller, Bruce. 2014. "An Ethnographic View of Legal Entanglements on the Salish Sea Borderlands." *University of British Columbia Law Review* 47 (3): 991–1023.

Mills, Antonia. 1994. *Eagle Down Is Our Law: Witsuwit'en Law, Feasts and Land Claims.* Vancouver: UBC Press.

Mills, Antonia. 2005. *"Hang onto These Words": Johnny David's Delgamuukw Testimony.* Toronto: University of Toronto Press.

Morales, Sarah Noël. 2014. "Snuw'uyulh: Fostering an Understanding of the Hul'qumi'num Legal Tradition." PhD diss., University of Victoria, Victoria, British Columbia. Accessed 29 Jan. 2017. https://dspace.library.uvic.ca/handle/1828/6106?show=full.

Morton, Ron, and Carl Gawboy. 2000. *Talking Rocks: Geology and 10,000 Years of Native American Tradition in the Lake Superior Region.* Minneapolis: University of Minnesota Press.

Nairn, Karen. 2005. "The Problems of Utilizing 'Direct Experience' in Geography
  Education." *Journal of Geography in Higher Education* 29 (2): 293–309. https://doi
  .org/10.1080/03098260500130635.
Noodin, Margaret. 2014. "Megwa Baabaamiiaayaayaang Dibaajomoyaang:
  Anishinaabe Literature as Memory in Motion." In *The Oxford Handbook of
  Indigenous American Literature*, ed. James H. Cox and Daniel Heath Justice,
  175–84. Oxford: Oxford University Press.
Overstall, Richard. 2005. "Encountering the Spirit in the Land: Property in a
  Kinship-Based Legal Order." In *Despotic Dominion: Property Rights in British
  Settler Societies*, ed. John McLaren, A.R. Buck, and Nancy E. Wright, 22–49.
  Vancouver: UBC Press.
Paige, S. Marlo. 2004. *In the Voices of the Sul-hween/Elders, on the Snuw'uyulh
  Teachings of Respect: Their Greatest Concerns Regarding Snuw'uyulh Today in the
  Coast Salish Hul'q'umi'num' Treaty Group Territory*. Unpublished MA thesis,
  University of Victoria, Victoria, BC.
Pardy, Bruce. 2010. "Ontario's Policy Framework for Environmental Education."
  Pathways: The Ontario Journal of Outdoor Education 22 (3): 22–3.
Pomedli, Michaeil. 2014. *Living with Animals: Ojibwe Spirit Powers*. Toronto:
  University of Toronto Press.
Potts, Gary. 1992. "Growing Together from the Earth." In *Nation to Nation:
  Aboriginal Sovereignty and the Future of Canada*, ed. Diane Engelstad and John
  Bird, 199–201. Toronto: House of Anansi Press.
Prosper, Kerry, L. Jane McMillan, Anthony A. Davis, and Morgan Moffitt. 2011.
  "Returning to Netukulimk: Mi'kmaq Cultural and Spiritual Connections with
  Resource Stewardship and Self-Governance." *International Indigenous Policy
  Journal* 2 (4): 1–17. https://doi.org/10.18584/iipj.2011.2.4.7 https://ir.lib
  .uwo.ca/cgi/viewcontent.cgi?referer=&httpsredir=1&article=1037&context
  =iipj. Accessed 24 Feb. 2017.
Quail, Susanna. 2014. "Yah'guudang: The Principle of Respect in the Haida
  Legal Tradition." *University of British Columbia Law Review* 47: 673–707.
Rajnovich, Grace. 1994. *Reading Rock Art: Interpreting the Indian Rock Paintings of
  the Canadian Shield*. Toronto: Natural Heritage/Natural History Inc.
Richland, Justin. 2008. *Arguing with Tradition: The Language of Law in Hopi
  Tribal Court*. Chicago: University of Chicago Press. https://doi.org/10.7208/
  chicago/9780226712963.001.0001.
Sable, Trudy, and Bernie Francis. 2012. *The Language of This Land, Mi'kma'ki*.
  Sydney, NS: Cape Breton University Press.
Sandy, Nancy. 2011. "Reviving Secwepemc Child Welfare Jurisdiction." In LL.M.
  thesis, University of Victoria,Victoria, British Columbia. https://dspace

.library.uvic.ca/bitstream/handle/1828/3336/Sandy_Nancy_LLM_2011
.pdf;sequence=1. Accessed 24 Feb. 2017.

Sarat, Austin, ed. 2004. *Law in the Liberal Arts*. Ithaca, NY: Cornell University Press.

Sarat, Austin, and Thomas R. Kearns, eds. 1995. *Law in Everyday Life*. Ann Arbor: University of Michigan Press.

Social Sciences and Humanities Research Council of Canada. 1983. Law and Learning / Le droit et le savoir: Report of the Consultative Group on Research and Education in Law. Ottawa: The Council. [Arthurs' Report.]

Truth and Reconciliation Commission of Canada. 2015. *Summary. Honouring the Truth, Reconciling for the Future*, vol. 1. Final Report of the Truth and Reconciliation Commission of Canada. Toronto: Lorimer.

Wa, Gisday, and Delgam Uukw. 1989. *The Spirit in the Land*. Gabriola, BC: Reflections.

Wildcat, Matthew, Mandee McDonald, Stephanie Irlbacher-Fox, and Glen Coulthard. 2014. "Learning from the Land: Indigenous Land Based Pedagogy and Decolonization." *Decolonization* 3: i–xv.

Wright, Walter. 1962. *Men of Medeek*. 2nd ed. Kitimat, BC: Northern Sentinel Press.

LIVE LIFE WITH SIGNIFICANCE ...

It was nine o'clock at night when our large bus pulled into the parking lot of a small town in rural Poland. I knew the name of the town at the time but have since forgotten it. We were all exhausted; we had been on the bus since 8:00 a.m., travelling around rural Poland seeing mass graves, cemeteries, and even a destroyed synagogue. It was our last stop, and all we knew was that it was another mass grave, a big one this time. We walked a little ways down a wooded area and arrived at the mass grave. A memorial had been set up. We all wandered from gravesite to gravesite on concrete paths as the sun went down, until the whole group congregated around one specific fenced-off area where, we were told, were buried hundreds of children, orphans, shot by Nazis during the Second World War.

On the fence around the grave, Jewish youths, as part of the March of the Living tour, had written notes to the murdered children. I bent down to read one: "I bet we are not that different." Never before in my life had I been slapped with so much perspective. It was true. I was not different from many of those children who were buried for 70 years now. The main difference between any one of those children and myself was that I got to grow up. My life was not cut short by senseless killing. The fact that I was tired, the fact that I was cold, none of it mattered. I got to be alive. Through nothing but chance I get to live out the rest of my life. The challenge now is to live it in a way that ascribes meaning to it, to honour those who had their lives cut short, and to live my life with significance because I am incredibly lucky to get to live it at all.

Freya Selander
Participant, University of Victoria's
I-witness Holocaust Field School

6

Putting Law in Its Place: Field School Explorations of Indigenous and Colonial Legal Geographies

DEBORAH CURRAN

Introduction: Place-less Law

Although the importance of place is foregrounded in field school education (Patel 2015; Poole and Hudgins 2014; Castleden, Daley, Morgan, and Sylvestre, this volume; Wright and Hodge 2012), the discipline and teaching of law is geographically uninterested (Blomley and Bakan 1992). At least in North America, almost all law school teaching takes place in a classroom and focuses on legal doctrine, the substantive principles that comprise the rules by which we agree to govern ourselves as citizens (Arthurs 2014; 2000). On a continuum of learning (Light, Cox, and Calkins 2009, 49), the vast majority of legal education would be stuck at the shallow learning end in the "recall" and "understanding" abilities, far away from "doing" or "changing." Experiential or skills-based learning in law is often limited to acquiring advocacy skills and providing legal advice through pro bono legal clinics or working for lawyers and acquiring skills pursuant to an apprenticeship model.

The study of law involves a seemingly infinite number of case studies. Each judicial decision that students read is based on a specific set of facts. However, the geographic context and embedded social relationships of these cases are usually filtered out. The nuance of a conflict within a community or between the state and individuals is often lost on students reading judgments based on decontextualized legal principles or by virtue of the sheer volume of cases they must consider each week.

After having taught the only ongoing Canadian interdisciplinary field school based in law for five years to a national law student body, I argue that field schools are an important methodology for teaching the complex application of law and problem solving in a specific legal context and are necessary for understanding Aboriginal and Indigenous law and the continuing conflict between Indigenous and colonial laws that underpins colonialism in Canada.[1] Colonial and Indigenous law – two components of the framework of law in Canada – are effectively taught "out there," away from a law school, in an explicit geographic context where place-specific relationships, issues, socio-ecological systems, political economies, and histories showcase law in action. The immersion of a field course experience, which includes hearing about relationships, conflicts, and expressions of jurisdiction from the individuals who live and work in that place, provides the needed context to assist students to find the place of law in a broader social, ecological, and economic context. It also injects place – a specific context – into law in furtherance of a just society.

At a basic level, well explored in other disciplines such as geography, a field school has the potential to offer legal education the "experiential, active learning that connects theory to practice, increases the enjoyment of, interest in, and affective connection to the subject and provides the opportunity to develop specific skills" (Owens, Sotoudehnia, and Erickson-McGee 2015, 314; Hope 2009). Moving beyond clinical legal education whose focus is often the acquisition of lawyering skills, field schools allow students to engage with the context of law and gain personal understanding of the way law interacts with and shapes relationships – interpersonal, political, and of power – within a specific community environment. As explicitly interdisciplinary, because the functioning of a community is interdisciplinary, learning in a field school format can get beyond the narrow learning of "the law" and provide much needed context for the most difficult part of becoming a legal practitioner: applying the law to specific circumstances in the furtherance of justice for a client or in the public interest.

It is at this most contextual and community-specific level that students can understand different types of law at work and how colonialism continues to shape socio-ecological experiences. Law schools teach colonial law. Yet, constitutionally affirmed and acknowledged Indigenous law is also alive and well across Canada (Borrows 2010). Coming from a colonial tradition, it is difficult for most law students to understand how community-specific Indigenous laws can operate alongside and challenge colonial law. Indeed, field courses can model how Indigenous law has shaped colonial law and vice versa given that Indigenous law arises from and reflects a specific geography (Asch, Borrows, and Tully 2018; Borrows, this volume). As called for by the Truth and Reconciliation

Commission of Canada (2015),[2] field schools can provide that necessary framework to explore legal pluralism and the enduring impact of colonialism. It is within this embedded contextual experience that most legal education subjects – family, municipal, environmental, debtor creditor, and international human rights law – can benefit from field school learning. Field school experiences are, however, particularly important for comprehending Indigenous law.

This chapter begins by highlighting common criticisms of legal education and locates experiential learning and field courses as a response, contending that Aboriginal and Indigenous law necessitate a place-based curriculum. Next, I describe the Hakai Field Course in Environmental Law and Sustainability offered by the University of Victoria Faculty of Law for five years, focusing on its learning outcomes, teaching methodology, and anonymous student evaluation. Students repeatedly reflected on the benefits of four related aspects of the field course experience – integration of course materials, immersion in the course experience, place-based learning, and the importance of relationships and complexity of issues. I discuss these four characteristics in the context of the criticisms of legal education and the call to participate in the process of reconciliation between settler and Indigenous communities in Canada – or decolonization.[3]

Legal Education and Learning

The problems with legal education in Canada and the United States, similar in strategy and structure (Arthurs 2000), are well documented: from its reproduction of illegitimate hierarchy and structures of power (Boyd 2005; Kennedy 2004), to suppression of emotion but yet its importance in intellectual thought (Harris and Shultz 1993), to graduates who do not gain appropriate skills to practise law (Sullivan et al. 2007; Cassidy 2012), to the failure to inculcate "civic professionalism" (Sullivan et al. 2007) or "humane professionalism" (Webb 2006). Legal education rapidly socializes students to "think like a lawyer" but does so in an acontextual way that relies primarily on one method of teaching, that of case-dialogue using analytical thinking to generalize, extract, and apply legal principles. This abstraction and simplification of personal and societal issues fails at "the task of connecting these conclusions with the rich complexity of actual situations that involve full-dimensional people, let alone the job of thinking through the social consequences or ethical aspects of the conclusions" (Sullivan et al. 2007, 6). At its worst, "the faith or dogma of law, its distance from subject, person or emotion, is precisely what precludes the dialogue or the attention to singularity which justice or ethics requires" (Goodrich 1996, vii).

In short, legal education teaches substantive law – the legal principles – but not the interpersonal and problem-solving skills to consider the law within the nuances of specific geographic, social, and cultural contexts. A stark example is most law students' exposure to Aboriginal law in a constitutional law course as a technical application of the test for infringement of Aboriginal rights divorced from the specific lands, waters, and Indigenous governance systems from which the jurisprudence arose. When considering whether a government-approved activity, such as logging or mining, in a specific watershed is an infringement of Aboriginal rights, the "takeaway" is a multipronged test that asks whether the Aboriginal right allegedly infringed is an integral practice, custom, or tradition of that Aboriginal society (*R* v. *Van der Peet* 1996) and whether the Crown adequately consulted and accommodated the Indigenous community (*Haida Nation* v. *British Columbia* 2004). Supreme Court of Canada Justice Thomas Cromwell (2010) has expressed similar sentiments in several fora, discussing how high quality, determinative, and reflective judgment is a product of evaluating complex facts, weighing multiple options, tailoring general knowledge to specific circumstances, taking many views into account, applying moral and ethical consideration by being empathetic but detached, and reflecting on and learning from experience.

Many point to clinical and experiential learning as a way to assist students with gaining "hands on" experience and application of substantive law.[4] Clinical legal education has typically focused on gaining lawyering skills by assisting clients who do not have access to legal services (Bloch 2008; Voyvodic 2001). Although it can deliver a more holistic experience, the intent of experiential education is broader and includes assisting students "to confront the normative, logistical, and relational issues that are immanent in all legal encounters" (Arthurs 2014, 713). These two pedagogies move towards more integrated legal education; however, they are often treated as add-ons rather than integral to a comprehensive educational experience (Sullivan et al. 2007).

More specifically related to geography, university- or urban-centred experiential learning is limited in its ability to teach legal disputes that are inherently place based or related to territory and geography. Much of environmental, Aboriginal, and Indigenous law is rooted in specific ecosystems and sociocultural contexts. Those contexts, and the inter-action between different jurisdictions, government actors, economic factors, and the larger Indigenous and non-Indigenous community in a watershed, is typically outside of law students' life experience. While they can learn the substantive principles of Aboriginal and environ-mental law, until they come into contact with Indigenous people's interaction with colonial law within their traditional territory, in the context of exercising even older unwritten Indigenous laws, students are unlikely to have a real understanding of the importance of place in Aboriginal and Indigenous law and how that embeddedness drives the approach to and resolution of specific legal issues. Indigenous law ema-nates, in part, from the land and water and shapes human behaviour and relationships (Asch, Borrows, and Tully 2018; Morales 2014). Like-wise, much of environmental law requires some fluency with the eco-logical function of a watershed and the impact of cumulative changes in that landscape.

Being in place is essential for having a deep understanding of Indig-enous and Aboriginal law and issues. As Borrows (1997) explains, "An Aboriginal consciousness is a spatial consciousness rather than a material consciousness" (453), noting the centrality of sharing space in Indig-enous cultures. Law is specific to that community and that place: "In these geographic spaces, [people] developed spiritual, political and social conventions to guide their relationships with each other, and with the natural environment. These customs and conventions became the foundation for many complex systems of government and law" (453–4).

Indeed, a primary technique of settler-colonialism has been to "sepa-rate Indigenous people from [their] sources of knowledge and strength – the land" (Wildcat et al. 2014, II) by deterritorializing activities and events. Reuniting people with the land is crucial for education to move towards decolonization within an Indigenous intellectual framework (Simpson 2014).

Over a short period of time, field courses can offer this opportunity to sit within a place and learn not just from the writing about that place, but also from the reciprocity between ecosystems, people, and governance processes. Field courses offer students the opportunity to experience law-in-action through the lens of a specific watershed or geography that creates place- and culture-specific nuances for interpreting and solving legal issues. This approach is particularly important for understanding Aboriginal and Indigenous law and the complex multijurisdictional and multisectoral nature of many disputes.

"Out There" Learning: The Hakai Field Course
in Environmental Law and Sustainability

From 2011 to 2015 the University of Victoria offered a Field Course in Environmental Law and Sustainability at the remote Hakai Institute, a field research facility in the Central Coast region of British Columbia. Located in the traditional territory of the Heiltsuk Nation and the Wuikinuxv Nation, the Hakai Institute is an hour and a half by boat from the nearest settlement. As a private research facility, the Hakai Institute hosts field courses primarily in scientific disciplines, including archaeology and geography. The field course was explicitly interdisciplinary and offered to students from both the faculties of Law and Social Sciences (through the School of Environmental Studies). The 2013 to 2015 courses were open to students from other law faculties across Canada. The instructor also offered spaces to students from both the Heiltsuk Nation and the Wuikinuxv Nation, with one student from the territory taking the course.

Between ten and seventeen students attended each year, for a total of sixty-three students (27 environmental studies and 36 law students, with an additional 10 teaching assistants equally drawn from the two disciplines). The field course occurred over twelve days at the Hakai Institute. In most years, the course started in Bella Bella, the Heiltsuk community where students met with staff members from the Heiltsuk Integrated Resources Management Department to discuss department work. The class travelled south together by boat through Heiltsuk and Wuikinuxv territory and then stayed at the Hakai Institute for the remainder of the course. Throughout the course, the students interacted with guests from the area, including staff of Fisheries and Oceans Canada, BC Parks, the Wuikinuxv Nation, non-governmental organizations, and community members and scientists conducting research from the Hakai Institute.

The course explored the complex structures of law and policy that shape environmental governance in the specific geography of the Central Coast region of British Columbia. The goal was to apply legal and policy lenses to the topics of (1) Aboriginal rights and title, (2) science in law, (3) land use and marine planning, and (4) the impact of energy systems on remote communities in order to help students understand the overlapping jurisdictional and governance systems that shape the region. The central themes that anchored the course were the geography of the Central Coast and the nearby Indigenous communities, ecological governance, ecosystem-based management, and herring as a vital ecosystem element.

The overarching teaching approach was that of a facilitated learning experience where the students contributed integrally to the evaluation of how environmental law and sustainability principles manifested in a specific place. The aim was to have what education theorists term a "deep" learning experience where students are encouraged to critique and reflect on the subject matter and their experience in a specific context (Wallace 2015). Students learned primarily from the wider classroom of the Central Coast, each other, guests, and the course materials. The instructor was the facilitator of this experience and assisted in setting the context, regulatory background, and governance structures within which students explored the topics addressed in the course. The classroom provided a framework from which students engaged with their physical environment, each other, and the guests, building on the information from the course materials.

Students were required to interact directly with the Central Coast as a place and the people who live and work within it. Rather than deal with a topic in environmental law or studies, the course explored how a region shapes Indigenous, Aboriginal, and colonial law and how these different forms of law shape the region with a view to expanding students' skills in applying law and policy analysis to current issues. Staff from First Nations in the region defined projects for which they needed research completed, and the intent was that students' final projects would deliver this research and be returned to the communities as a contribution of the course to the knowledge about the Central Coast.

The students moved beyond the typical law class approach of discussing the elements of the test for Aboriginal rights, a colonial construct under the Constitution Act, 1982, and its jurisprudence (*Haida Nation* v. *British Columbia* 2004; *Tsilhqot'in Nation* v. *British Columbia* 2014). Students interacted with the manifestation of what is required within Crown-Indigenous consultation and accommodation as specific on-the-ground or in-the-water examples explained by both Indigenous and Crown actors. The test for Aboriginal rights, and its place-based specificity, came alive as the class walked through a 5000-year-old midden or kayaked adjacent to clam gardens. In addition, in several of the courses students experienced acts of Indigenous law by being invited to witness ceremonies or dances. They also heard from staff from First Nations about fish enhancement and documentation programs, as well as community fisheries, which are modern expressions of Indigenous law.

The students completed anonymous course experience surveys.[5] These evaluations are consistent with the literature on field courses

(Ward 1999; Fuller et al. 2006; Mitussis and Sheehan 2013), where students revealed a high degree of satisfaction with the course and regularly reflected on their "deep learning," remarking that the course "changed the way I think about things" and was "the most effective learning experience of my post-secondary education" (Anonymous students, pers. comm.). These reflections describe some results of the course experience, but they do not reveal specifically what pedagogical aspect of this field course precipitated the students' learning. More specifically, the student survey responses emphasize four recurring characteristics of the field course that can help explain the students' immediate sense of valuable learning: (1) the relational way the course topics integrated with one another; (2) the benefits of immersion in a course; (3) being in the place from which the issues, policy, and law emanate and within which they are applied; (4) and interacting with people who are experiencing and making law and/or policy, thereby being confronted with the complexity of the issues.

All of these course elements, discussed below using quotations from the student evaluations, intermingle to present a more nuanced account of the law employed in a complex and specific context. This approach requires students to think about the application of the law from different perspectives, and to use extra-legal skills to solve large-scale issues.

Integration

When studying law in a specific place it is impossible to get away from the interconnection of issues in that watershed community. Each academic topic builds on the geographic, social, and political context, and these all relate to one another. Each course unit offers students the opportunity to expand their existing knowledge and reconsider previous topics in light of new information as "everything was related in complex and interesting ways" (Anonymous student, pers. comm.).

This learning experience is starkly in contrast to typical classroom-based learning in law:

> The strength of the course was to have these overarching themes that we touched on all the way through the course. In black letter law courses we jump from thing to thing and the topics don't necessarily hang together. The way this course was, it was more realistic about how the law really is. The repetition helped me to learn the layers. It felt more fulsome. (Anonymous student, pers. comm.)

An example of this integration was the focus throughout the course on herring, an essential actor in the Central Coast ecosystem and cultures. Herring plays a foundational role in Aboriginal rights jurisprudence; the *R* v. *Gladstone* decision, involving individuals from the Heiltsuk Nation and the sale of herring roe-on-kelp, was the first acknowledgment of a commercial right to fish (1996). Herring is also central to the discussion of science and law and marine use planning where the scientific uncertainty over herring stocks leads to contrary political decisions and then challenges in court (*Ahousaht First Nation* v. *Canada* 2014). The fisheries officers also talk about the herring fishery from an enforcement perspective. During the field school, the topic of herring is just one example of the interconnections of many diverse threads of inquiry. These examples cover several different topic areas of law. When examined as part of the legal geography of the Central Coast, the context and application to that context – like all legal dispute resolution – is what is important. Students gain a more integrated and fulsome account of the law by examining legal processes through the lens of an important community icon.

This integration of topics and materials allows students to reflect on their learning continually through the course and thus more fully embed knowledge. This responds to the students' need for "a dynamic curriculum that moves them back and forth between understanding and enactment, experience and analysis" (Sullivan et al. 2007, 8). This approach also realizes interdisciplinarity both across legal topics and in learning the application of the law. One cannot contemplate "herring law" without considering the cultural, ecological, resource management, and political contexts in which it is embedded. For legal education, interdisciplinarity is necessary as an "essential element of the response to change" (Arthurs 2014).

Immersion

Immersion relates both to becoming intimate with a physical place (addressed below under place-based learning) and the committed nature of the field course itself, where students engage with the course topics for most of their waking hours. The field course requires students to live, work, and recreate in the place about which they are learning. Students are not passive recipients of words in the classroom from an instructor or guest. They are not limited to a one-and-a-half-hour classroom segment to think about a specific legal principle. They use all of their senses to incorporate the materials into their understanding. Through various intellectual and physical activities they reflect on the course materials in different ways and at different times. Outside the classroom they are interacting with one another during meals, on hikes, and in their rooms or the common space. They are also interacting with the land and sea-scape that is the subject of the law they are studying: students are embedded in the landscape. This physical immersion, uninterrupted by the usual concerns with daily living and social responsibilities, stimulates in unstructured and serendipitous ways ongoing thinking about the course topics that is not just about what the students are learning, but also about the future importance of the place and the issues important to it. They are an intrinsic part of the landscape of learning.

Knowledge generation occurs continuously as the course experiences stimulate intellectual, physical, and emotional responses:

> The act of being here, talking with everyone, and being immersed in what you are learning [is a] full body experience. (Anonymous student, pers. comm.)

> I will remember a lot more about this course and the materials than sitting in a lecture hall. (Anonymous student, pers. comm.)

One sees the benefits of immersion most forcefully when a classroom experience with community guests is followed by a meal or walk to the beach where the guest extemporaneously shares more of their knowledge and lived experience with the students. One student recounted the joy and honour of canoeing with an Elder to gather eelgrass for the Elder to take back to her community. This intensity of the field learning experience allows students to be "immersed in their learning, rather than confronted by it at intervals in the lecture room or during assessment" (Mitussis and Sheehan 2013, 52).

Place-Based Learning

The physical nature of being in the place that is the teaching subject and physically immersed allows learning to occur in many different ways. Students do not just read about issues or law, they experience them as important and contemporary concerns in the environment and for the individuals within the community:

> Being here at Hakai, studying Aboriginal and environmental law was extremely important in helping me to learn and interact with the course materials. (Anonymous student, pers. comm.)

Studying environmental law in a classroom is about legal principles, but studying it in a particular watershed offers the opportunity for students to see the law emanating from a tree, a fishing vessel, or from a piece of kelp. The class studies the structures of law and environmental management that respond to a specific ecosystem element and that shape those elements in particular ways:

> You get to experience the law in a way that law students never experience. When you are on the dock and you experience different organisms in each different place – that is the way the law works. (Anonymous student, pers. comm.)

> The transformative part was going bottom up with law – taking the law from the place where we are and to see how law is displayed here. (Anonymous student, pers. comm.)

An important example of a transformative "taking the law from this place" was witnessing the reopening of a camp involving a feast, speeches, and acknowledgment of family and neighbours. Neighbouring First Nations undertook songs and dances that were enactments of Indigenous law. The recounting of territory and relationships through dance and song was Indigenous governance. It was only by being present at that event that students could achieve some insight into what had been, up until that moment, disassociated discussions of Indigenous law, Aboriginal law, and colonial governance in that specific place. This multisensory and ongoing interaction within the course moves participants beyond talking about land or a watershed, the subject, to engaging with the land and understanding it in multiple ways (Wildcat et al. 2014; Aldern and Goode 2014).

Relationships and Complexity

The other aspects of field schools would be less meaningful without interacting with the citizens of the landscape in which the learning is taking place. Hearing from the individuals who are involved in the law and whom the law affects is a powerful experience as "you really learn how law works on the ground or how it plays out when applied" (Anonymous student, pers. comm.). While we can do that in the classroom, the embodied nature of being in the place where these issues are occurring resonates with the students and brings the material alive:

> Relationships with First Nations allowed students to learn directly from Indigenous communities [and] challenges students to think critically about environmental and Aboriginal law. (Anonymous student, pers. comm.)

While care must be taken not to burden local community members with participation, the "visitor and visited" share and produce knowledge together (Vibert and Sadeghi-Yekta, this volume). This connection between individuals from different communities and experience provides context and nuance to complex legal, social, and political issues. Face-to-face interaction fosters an appreciation of different points of view, and ideally, understanding and an ability to think through multiple pathways for solutions. Hearing a staff member of a First Nation talk about colonial fisheries mismanagement is contrasted with fisheries officers unpacking the challenges with enforcement in a remote area of the coast. This is law school at its most comprehensive, as "knowledge communities" provide "students with context and perspective that allow them to make sense of what they have observed or experienced" (Arthurs 2014, 713).

Gaining an appreciation for this complexity, both legal and ecological, is enhanced by learning in place:

> It is not possible to think usefully about legal education without thinking about law itself – the subject that is taught – about its complexity and polycentricity, its political and economic functions in the larger society, its social origins and cultural significance, its epistemology and deontology. Nor is it possible to think about the architects, theorists, practitioners, critics, clienteles, benefactors, and beneficiaries of legal education without recalling that they are also embedded in the larger polity, society, economy, culture, professional ethos, and higher education system – all dynamic and conflicted systems. (Arthurs 2000, 402–3)

Re-placing Law

Returning to the criticisms of legal education and its one-size-fits-all approach to learning to "think like a lawyer" that focuses overwhelmingly on disembodied case analysis in the classroom, field courses address some of the most challenging aspects of developing "civic professionalism" (Sullivan et al. 2007) or "humane professionalism" (Arthurs Report 1983). Learning to behave as a humane professional participating in the mandate for decolonization, as emphasized by the Truth and Reconciliation Commission of Canada, requires an ability to understand the nuanced and complex context of issues and the role that law plays while seeking creative solutions. Field courses can provide this integrated and multifaceted approach to course content, both legal and non-legal, in the place that is the subject matter of the content while gaining context for the application of the law and the complex issues from people who live and work in that place.

Before unduly constraining the application of geography only to environmental, Aboriginal, or Indigenous law, I note that field courses can be applicable to a wide range of legal subjects; further, I argue that geography is important for all kinds of law. For example, a criminal law case relating to the confidentiality of documents and protection of witnesses can be embedded in the context of abuse at a specific residential school. The roots of other criminal and child protection cases are often traceable back to the legacy of colonialism. At the same time as the criminal activity is occurring, First Nations are asserting their rights to the land and waters, challenging Crown decisions using administrative law and making declarations of Indigenous law (Syilx Nation 2014; Curran 2015; Nadleh Whut'en 2016). This nuanced context of a specific place helps explain choices about and the impact of using legal remedies (or not) and fosters integrated, complex, problem solving.

As field schools are taken up, consistent with the decolonization imperative, educators must pay close attention to the need to "diversify representation by attending to power relations and drawing on a range of local voices and experiences" (Vibert and Sadeghi-Yekta, this volume). Finally, putting law in its place is a prerequisite to understanding much of Indigenous and Aboriginal law, and participating in, at minimum, the reconciliation between Indigenous peoples and settler society. As revealed by Indigenous scholars, Indigenous laws and culture are specific to a geographic place within which social, political, and ecological relationships interact to shape and be shaped by law. It is in this space that field courses can offer a venue through which students can experience the nuanced context of the complexity of legal pluralism and multifaceted problem solving.

NOTES

Special thanks to the Hakai Institute, Christina Monk, and Eric Peterson who provided foundational support for the field school described in this chapter.

1 State- or Crown-initiated law of settler societies predominates in countries with a colonial past such as Canada, the United States, and Australia.

148 Deborah Curran

Most people think of "the law" as emanating from an elected centralized government and courts. However, there are typically many types of law at work. We use common law in most of Canada and civil law in Quebec as colonial law, but each Indigenous community has its own laws, and many people are subject to religious laws. Aboriginal law refers to section 35 of the Constitution Act, 1982 in Canada that acknowledges and affirms Aboriginal rights and the jurisprudence that defines those rights.

2 Specifically, the Truth and Reconciliation Commission of Canada, in Calls to Action 27 and 28, calls on law societies and law schools to provide lawyers and law students with training in Aboriginal people and the law, the history and legacy of residential schools, treaties, Aboriginal and Indigenous law, Aboriginal-Crown relations, intercultural competency, conflict resolution, and human rights.

3 Many Indigenous scholars point to decolonization, a moving away from reconciliation between Aboriginal peoples and the Crown to an Indigenous intellectual and governance paradigm. See, e.g., Alfred (2009) and Coulthard (2014). While I use both "reconciliation" and "decolonization" as terms in this chapter, in law, governance, and practice, they are different frameworks.

4 Clinical legal education provides the opportunity for law students to gain legal skills by providing legal information and performing permitted lawyering tasks through a legal aid clinic or service. Experiential education, in contrast but not in exclusion to clinical legal education, should have three elements: (1) a managed learning process, (2) a planned experience, and (3) reflection, evaluation, and/or feedback: "Education, in contrast to a learning opportunity, consists of a designed, managed and guided experience" (Moliterno 1996, 78).

5 Course experience surveys are administered and collated by the university. The instructor is not present when the students complete them, and the university provides the anonymous collated results to the instructor after the course is completed. While most of the questions are scored on a scale of 1 to 5, there is space for students to provide their suggestions for improving the course and to describe aspects of the course that assisted them with their learning in the course.

REFERENCES

Ahousaht First Nation v. Canada (Fisheries and Oceans). 2014. FC 197.
Aldern, Jared Dahl, and Ron W. Goode. 2014. "The Stories Hold Water: Learning and Burning in North Fork Mono Homelands." Decolonization 3 (3): 26–51.

Alfred, Taiaiake. 2009. *Wasáse: Indigenous Pathways of Action and Freedom.* Toronto: University of Toronto Press.

Arthurs, Harry W. 2000. "Poor Legal Education: So Near to Wall Street, So Far from God." *Osgoode Hall Law Review* 38 (3): 381–408.

Arthurs, Harry W. 2014. "The Future of Law School: Three Visions and a Prediction." *Alberta Law Review* 51 (4): 705–16.

Arthurs Report. See Social Sciences and Humanities Research Council of Canada.

Asch, Michael, John Borrows, and Jim Tully, eds. 2018. *Resurgence and Reconciliation: Indigenous-Settler Relations and Earth Teachings.* Toronto: University of Toronto Press.

Bloch, Frank S. 2008. "Access to Justice and the Global Clinical Movement: New Directions in Clinical Legal Education." *Washington University Journal of Law and Policy* 28: 111–40.

Blomley, Nicholas K., and Joel C. Bakan. 1992. "Spacing Out: Towards a Critical Geography of Law." *Osgoode Hall Law Journal* 30:661–90.

Borrows, John. 1997. "Living Between Water and Rocks: First Nations, Environmental Planning and Democracy." *University of Toronto Law Journal* 47 (4): 417–68. https://doi.org/10.2307/825948.

Borrows, John. 2010. *Canada's Indigenous Constitution.* Toronto: University of Toronto Press.

Boyd, Susan B. 2005. "Corporatism and Legal Education in Canada." *Social & Legal Studies* 14 (2): 287–97. https://doi.org/10.1177/0964663905051225.

Cassidy, R. Michael. 2012. "Beyond Practical Skills: Nine Steps for Improving Legal Education Now." *Boston College Law Review* 53 (4): 1515–32.

The Constitution Act, 1982, Schedule B to the Canada Act 1982 (UK), 1982, c. 11 s. 35.

Coulthard, Glen Sean. 2014. *Red Skins, White Masks.* Minneapolis: University of Minnesota Press. https://doi.org/10.5749/minnesota/9780816679645.001.0001.

Cromwell, Thomas. 2010. Keynote Address. Conference on Canadian Clinical Legal Education, University of Western Ontario. 22 Oct.

Curran, Deborah. 2015. "Water Law as Watershed Endeavour: Federal Inactivity as Opportunity for Local Initiative." *Journal of Environmental Law and Practice* 28 (1): 53–88.

Fuller, Ian C., Sally E. Edmondson, Derek France, David Higgitt, and Ilkka Ratinen. 2006. "International Perspectives on the Effectiveness of Geography Fieldwork for Learning." *Journal of Geography in Higher Education* 30 (1): 89–101. https://doi.org/10.1080/03098260500499667.

Goodrich, Peter. 1996. *Law in the Courts of Love: Literature and Other Minor Jurisprudences.* New York: Routledge.

Haida Nation v. *British Columbia* (Minister of Forests), [2004] S.C.R. 511.

Harris, Angela P., and Marjorie M. Shultz. 1993. "'A(nother) Critique of Pure Reason': Towards Civic Virtue in Legal Education." *Stanford Law Review* 45 (6): 1773–1805. https://doi.org/10.2307/1229127.

Hope, Max. 2009. "The Importance of Direct Experience: A Philosophical Defence of Fieldwork in Human Geography." *Journal of Geography in Higher Education* 33 (2): 169–82. https://doi.org/10.1080/03098260802276698.

Kennedy, Duncan. 2004. *Legal Education and the Reproduction of Hierarchy: A Polemic Against the System.* New York: New York University Press.

Light, Greg, Roy Cox, and Susanna Calkins. 2009. *Learning and Teaching in Higher Education: The Reflective Professional.* 2nd ed. Los Angeles: Sage.

Mitussis, Darryn, and Jackie Sheehan. 2013. "Reflections on the Pedagogy of International Field-schools: Experiential Learning and Emotional Engagement." *Enhancing Learning in the Social Sciences* 5 (3): 41–54. https://doi.org/10.11120/elss.2013.00013.

Moliterno, James E. 1996. "Legal Education, Experimental Education, and Professional Responsibility." *William and Mary Law Review* 38: 71–123.

Morales, Sarah Noël. 2014. "Snuw'uyulh: Fostering an Understanding of the Hul'qumi'num Legal Tradition." PhD diss., University of Victoria, Victoria, British Columbia. Accessed 29 Jan. 2017. https://dspace.library.uvic.ca//handle/1828/6106.

Nadleh Whut'en and the Stellat'en First Nations. 2016. Yinka Dene 'Uza'hné Surface Water Management Policy. Version 4.1. 18 March. Accessed 29 Jan. 2017. http://www.carriersekani.ca/images/docs/Yinka%20Dene%20'Uzah'ne%20Surface%20Water%20Management%20Policy%20%28March%2018%20 2016%29%20%2800303183xC6E53%29.pdf.

Owens, Cam, Maral Sotoudehnia, and Paige Erickson-McGee. 2015. "Reflections on Teaching and Learning for Sustainability from the Cascadia Sustainability Field School." *Journal of Geography in Higher Education* 39 (3): 313–27. https://doi.org/10.1080/03098265.2015.1038701.

Patel, Kamna. 2015. "Teaching and Learning in the Tropics: An Epistemic Exploration of 'The Field' in a Development Studies Field Trip." *Journal of Geography in Higher Education* 39 (4): 584–94. https://doi.org/10.1080/03098 265.2015.1084499.

Poole, Amanda, and Anastasia Hudgins. 2014. "'I Care More About This Place, Because I Fought for It': Exploring the Political Ecology of Fracking in an Ethnographic Field School." *Journal of Environmental Studies and Sciences* 4 (1): 37–46. https://doi.org/10.1007/s13412-013-0148-6.

R v. *Gladstone,* [1996] 2 S.C.R. 273.

R v. *Van der Peet,* [1996] 2 S.C.R. 507.

Simpson, Leanne Betasamosake. 2014. "Land as Pedagogy: Nishnaabeg
Intelligence and Rebellious Transformation, Decolonization: Indigeneity."
Education and Society 3 (3): 1–25.
Social Sciences and Humanities Research Council of Canada. 1983. Law
and Learning / Le droit et le savoir: Report of the Consultative Group on
Research and Education in Law. Ottawa: The Council. [Arthurs Report.]
Sullivan, William M., Anne Colby, Judith Welch Wegner, Lloyd Bond, and
Lee S. Shulman. 2007. *Educating Lawyers: Preparing for the Profession of Law.*
San Francisco: Carnegie Foundation for the Advancement of Teaching and
Jossey-Bass.
Syilx Nation Siwłkʷ Declaration. 2014. Declared at the Okanagan Nation
Alliance Annual General Assembly 31 July 2014. Accessed 29 Jan. 2017.
https://www.syilx.org/wordpress/wp-content/uploads/2012/11/Okanagan
-Nation-Water-Declaration_Final_CEC_Adopted_July_31_2014.pdf.
Truth and Reconciliation Commission of Canada. 2015. *Summary. Honouring
the Truth, Reconciling for the Future*, vol. 1. Final Report of the Truth and
Reconciliation Commission of Canada. Toronto: Lorimer.
Tsilhqot'in Nation v. *British Columbia*, 2014 SCC 14.
Voyvodic, Rose. 2001. "Considerable Promise and Troublesome Aspects: Theory
and Methodology of Clinical Legal Education." *Windsor Yearbook of Access to
Justice* 20: 111–40.
Wallace, Susan, ed. 2015. *A Dictionary of Education.* 2nd ed. Oxford: Oxford
University Press.
Ward, Martha C. 1999. "Managing Student Culture and Culture Shock: A Case
from European Tirol." *Anthropology & Education Quarterly* 30 (2): 228–37.
https://doi.org/10.1525/aeq.1999.30.2.228.
Webb, Julian. 2006. "The 'Ambitious Modesty' of Harry Arthurs' Humane
Professionalism." *Osgoode Hall Law Journal* 44 (1): 119–55. http://
digitalcommons.osgoode.yorku.ca/ohlj/vol44/iss1/6.
Wildcat, Matthew, MandeeMcDonald, StephanieIrlbacher-Fox, and
GlenCoulthard. 2014. "Learning From the Land: Indigenous Land-Based
Pedagogy and Decolonization." *Decolonization* 3 (3): I–XV.
Wright, Sarah, and Paul Hodge. 2012. "To Be Transformed: Emotions in Cross-
Cultural, Field-Based Learning in Northern Australia." *Journal of Geography
in Higher Education* 36 (3): 355–68. https://doi.org/10.1080/03098265.2011
.638708.

MYSTERIES REMAIN ...

It is our first day taking the bus in India. The buses are crowded, hot, and sweaty. Imagine how much fun taking the bus is in Canada and then make it less fun but somewhat more interesting, and that pretty much sums up what it's like to take the bus in India.

Taking the bus to Pondicherry is fairly straightforward. You just have to stand on the side of the road and wave your arms for a while, and eventually a bus driver will take pity on the poor foreigner and slow down long enough for you to jump in (the buses have no doors. Who really needs doors anyway?). Then, once you have clambered onto a moving bus, you take a crowded, hot, sweaty ride to Pondicherry, where you get dropped off at the bus station, which is a bit chaotic, but at least you know where you are. There is a mass exodus of humans from the bus, and it is clear that it is time to remove yourself from the vehicle.

Taking the bus back to Tamaraikulam, however, is a different story. You have to get on the right bus and then eventually communicate to the whistle-blowing man in the brown uniform that you need to get off at RKITI. What is RKITI? This is one of the mysteries of the universe.

It is also worth mentioning that every person in India who has a job seems to wear the same brown uniform. Bus drivers, auto rickshaw drivers, police officers, security guards, and so on.

So, on our first day taking the bus in India, we have trouble figuring out what bus we're supposed to get on. One of the major problems is that our desired bus is the bus to Cuddalore, a place no foreigners ever go because it is the touristic equivalent of going to Humboldt, Saskatchewan. So, when we say "bus to Cuddalore?" people often either stare blankly or ask "Cuddalore? Why?"

Finally, we find a man who helps us. He even writes down the name of our stop in Tamil. He ushers us onto the bus and makes absolutely sure that it is the right one.

Two months later, we are standing at a different bus stop. A man says, "Hello! I remember you. I helped you get on a bus to Cuddalore! RKITI!"

Sure enough, it is the same fellow. He is pleased to see that we now understand the Indian bus system on a much deeper level than we previously did.

RKITI remains a mystery.

Laura Buchan
Participant, University of Victoria's
Applied Theatre Field School in India

7

Power in Place: Dilemmas in Leading Field Schools to the Global South

ELIZABETH VIBERT AND
KIRSTEN SADEGHI-YEKTA

Introduction

Short-term study visits to the Global South are laden with tensions. Among the most obvious are tensions between the desire for the exotic that may motivate students to go to such "out there" places in the first instance and the critical imperative to narrow the distance between South and North; tensions between the privilege of students able to set off on costly international trips and the poverty that characterizes many of the communities visited; and tensions between a commitment to social justice objectives (often on the part of instructors and students alike) and the deleterious effects of such travel on the environment and, potentially, on host communities. As practitioners from history and theatre – disciplines with little tradition of field study – we approach field study programs with a mixture of enthusiasm and scepticism. In this chapter we focus on several aspects of instructor power and positionality that warrant consideration when planning and delivering a field school program. These considerations are relevant to any field school, but especially those in colonized spaces. With Global South settings in mind, we focus here on the ethical dilemmas of defining and entering a community; the challenge of deploying the analytical value of "difference" while working to shine light on similarities and conjunctures between South and North; the dangers of scratching the surface on a short-term visit; and the risk of inducing a sense of futility in well-intentioned, critically engaged learners. Finally, in partial response to these challenges and dilemmas, we consider practical ways to ensure benefit (and certainly not harm) to the communities visited. As a social historian and an applied theatre scholar we are more concerned with evidence and praxis than with abstract theory; the theoretical underpinnings of our approach will be clear at various points.

As students of colonialism, we are acutely aware of the long and often disastrous history of metropolitan travellers making sport, dispensing philanthropy, collecting specimens, and otherwise consuming the landscapes and peoples of colonized spaces. Colonial regions of the era of European empires prefigure the Global South or "tropical" spaces of the twenty-first century field school.[1] Scholars of colonial representation (Bhabha 1994; Said 2003, 1993; Spivak 1988; Wheeler 2000) emphasize the ways "Europe" or "the West" imagined and created the tropics as profoundly other to temperate regions – impoverished, disordered, and diseased as against the latter's modernity, civilization, and enlightenment. Arnold highlights the paradox in the way tropical spaces or spaces of the South are today seen as at once "natural spectacle" arousing curiosity (Patel 2015, 587; Arnold 2000) and deficient subjects of pity or disdain. Postcolonial theory and feminist methodologies have nudged scholars in history, theatre, and many other disciplines to reflect on their own investments in such representations, to conscientiously engage local knowledges, and to seek to understand places in local terms (Geertz 1983; hooks 1989; Alexander and Mohanty 1997; Appadurai 2000; Mignolo 2000; Lugones 2007; Newstead 2009; Clifford 2013). In the memorable words of cultural theorist Stuart Hall, "It is when a discourse forgets that it is placed that it tries to speak for everybody else" (1997, 36). Short-term study visits to the Global South, with their packed agendas and frequent logistical challenges, run the risk of forgetting how they are "placed": as Patel argues, the impulse to decolonize and diversify representation by attending to power relations and drawing on a range of local voices and experiences is often "missing in the thinking and production of field trips" (2015, 587).

This chapter ventures into that gap. Where Patel focuses on ways to ensure students understand themselves as subjects in the field, tasked with reflecting on and coming to know their own subjectivities, here we consider the positionality of the instructor leading the trip. How can instructors designing and delivering field schools create the conditions for participants – ourselves included – to reveal scholarly, political, cultural, and personal investments and to interrogate assumptions about peoples and places "out there"? How can instructors ensure ethical attention to the range of local knowledges and experiences at play in the communities we visit? How do field school faculty manage and manipulate *power in place*, and how can we do so in ways that respect the subjectivities of local people and of the students in our care? This chapter does not attempt to answer these questions so much as to shine light on the kinds of dilemmas facing an engaged critical pedagogy in "the field" (Patel 2015) of the Global South.

The Field Schools

Our field schools are based in rural communities in South Africa and a small rural town in Nicaragua. The Colonial Legacies Field School in South Africa accepts approximately fifteen senior undergraduates, most from the University of Victoria in British Columbia, Canada. After a week of classroom and preparatory work, the group travels to South Africa, spending three weeks on the ground, two based in the villages of a communal territory in Limpopo Province. Students participate in "everyday" activities ranging from vegetable farming to teaching to craft production (as volunteers) through water collection and cooking (as guests). They practise the techniques of oral history, collecting data for research essays and public presentations to be completed on their return to Canada. As the name "Colonial Legacies" suggests, the field school aims to expose students to the contemporary legacies of the country's multiple colonial pasts. We move about the country animated by one overarching question: how are the pressures and processes of South Africa's complex colonial histories written on the landscape and on people's lives today? This question, attentive as it is to both the patterns of the past and present-day challenges – to the way the past is alive in the present – compels us to think in registers both global and profoundly local. How did British colonial tax laws of the late nineteenth century contribute to the depressed state of small-scale agriculture today? How have migrant labour systems – enforced first by global industrial capital, then by the apartheid regime, and more recently, by structural unemployment – shaped family relations into the present? Global patterns are laid out primarily through reading, but attention to local voices brings those patterns into sharp relief. On a visit to a citrus packing house belonging to a global corporation, for instance, the manager explains that most of the fruit is exported, and local people tell us they rarely have a chance to taste premium fruit, despite living in *sitrus wereld* (Afrikaans for citrus world).

The Applied Theatre Field School in Nicaragua will accept between ten and fifteen students, upper-level undergraduates from the University of Victoria. Prior to travel, students will take a one-term, on-campus course called "Theatre, Conflict and Development." In addition to academic work during the course, students also practically prepare for the field school: through drama workshops and via Skype meetings with theatre practitioners from the Global South, they learn about cultural awareness, ethics, safety, and translation. The field school itself is based in La Chispa, a small rural town close to Matagalpa in Northern Nicaragua. The school runs two to three weeks. Students have the opportunity to visit a range of applied theatre companies in the community. The companies work with a variety of client groups such as children, young adults, and farmers. The main aim of the Applied Theatre Field School is to offer a practical exploration of theatre in developing contexts and (post)-conflict zones supported by an analysis of the value of theatre practitioners working in these settings. At the same time, the field course explores some of the fundamental concerns around the potential of the arts in marginal-ized settings. The course assists students in approaching applied theatre critically and analytically with regards to social justice, as well as to issues relating to theatre production and patronage, the geopolitical diffusion of style and repertoire, globalization, and post-conflict and development contexts. The Applied Theatre Field School attempts to create awareness of the important place applied theatre occupies within marginalized and globalized cultures and societies.

As leaders of community-engaged field schools, we proceed from the conviction that for pragmatic and political reasons, it is essential to have established prior meaningful connections with the community members and groups we visit (Broekman 2014; Menzies and Butler 2011). Beyond positive personal relationships, the village-rootedness of our field schools requires community permission and a degree of local enthusiasm. It is no small undertaking for resource-poor com-munities to host a sizeable group of international visitors for 10 days or more. However, prior community relationships cannot sidestep the potential for "domination" – for "try[ing] to speak for everybody else," in Hall's words (1997, 36). The fact that Kirsten's theatre networks in Nicaragua and Elizabeth's networks in South Africa spread according to our long-term research contacts with artists, farmers, and traditional authorities means that our field schools are permeated by the observa-tions and perspectives of those particular groups. Moreover, our con-cern about "domination" reflects our understanding that *we* as visitors and researchers are not *they* in the community, nor are we the subjects of research. From the start the relationship is hierarchical; the field schools are charged with unequal power relations in ways we explore below (LaSalle 2010). Hierarchies are also clearly at play between instructors and students. A stark example is Kirsten's decision to cancel her scheduled field school to Nicaragua in 2017 due to the outbreak of Zika virus. With the knowledge that many theatre students had been counting on participating in this field school and would graduate without doing so, Kirsten exercised her ultimate power as instructor: she took this opportunity away from the students.

Power in Place: Ethical Dilemmas

What Community?

Here is the first problem for community-engaged field schools: what is the "community" in which we are based? And what does "community based" mean when the visit is short term (less than a month; Glass 2015)? We are certainly not there long enough to integrate into the community. Are we there long enough for this new context to become our local? Becoming local, however briefly, is often held up by self-styled travellers as a virtue of their approach, in contrast to the tourist mode of ticking off sites on a top-10 list (Urry 2002; McMorran 2015). South Africa's particular history of state-driven racial discrimination, and its enduring racialized class fissures, make fitting into a village a challenge for visitors identified as white. In Nicaragua, too, short-term visitors have to be prepared to stand out – a significant experience for those accustomed to the privileges of unmarked whiteness. But if we cannot become local, field school visitors can at least aspire to comprehend a place in local terms – not defined by sameness or difference, but as a "contextually specific space of betweenness" (Newstead 2009, 81). Here, visitor and visited "actively meet and, through always uneven dialogue and negotiation, produce strong, yet situated and partial, knowledge" (Newstead 2009, 81; see also Bakhtin 1975; Clifford 2013; hooks 1997).

"Community" is an essentially contested concept; it can mean almost anything one wishes it to mean (Garver 1978; Hoggett 1997; Li 1996). We grapple with the meaning of the concept in planning our field schools and the questioning continues long after the fact. For the purposes of this chapter, we take community to be comprised of three elements – place, memory, and interaction (Anderson 1983; Bauman 2000; Booth Fowler 1991; Davidoff et al. 1999; Giddens 1984; Kepe 1999; Kymlicka 1995). First, communities are often linked to a locality, a physical space. There is frequently an element of affect to this linkage, in the sense that the place is conceived of as "home." Community as place has a troubled history in South Africa, where over the years millions of people were forcibly relocated on the basis of ascribed racial and ethnic identities. Similarly, after a history of colonization, Nicaragua has suffered from natural disasters, civil wars, and dictatorships (Federal Research Division 1999). Nonetheless, Westlake argues that the Nicaraguan population has always articulated inclusion primarily through *mestizaje* (mixing), that is, "where Europeans create nationalist sentiments out of constructions of 'racial purity'" (2005, 44); by contrast the unification of Nicaraguans comes from solidarity between racialized peoples. Memory functions in these settings in potent ways. The people of the villages in Limpopo, South Africa, for instance, share both a deep memory of the peopling of the area by their ancestors (an element regularly rehearsed by the traditional leadership) and a more recent and negative memory of being pressed into cramped bureaucratic space by the functionaries of apartheid (Hay 2014; Harries 1989). As leaders of the field schools, we try to ensure students have read about these complicated and powerful associations before arriving on the ground.

The third sense of community, based in face-to-face interaction, is both the most meaningful in daily life and the most exclusionary. This is an element of community Elizabeth seeks to emphasize in the South Africa field school, partly for the way it draws attention to the atomistic tendencies of our own society, situating and historicizing the personal habits of neoliberalism (Evans and Sewell 2013; Hall and Lamont 2013). Elizabeth and the students work alongside women farmers who have nurtured a community of solidarity born out of their shared experiences of famine and gendered impoverishment. Participants work alongside youth farmers who have created a community of mutual trust and support through their shared experience of learning to farm and working together every day in the dust and heat (Vibert 2016). Elizabeth valorizes these groups without apology. She accepts their narratives of mutualism and service of a greater good. Their stories resonate with her own ethical stances towards food and community, and what she likes to consider her alienation from neoliberal imperatives. But what do the instructor's politics mean for the host community and for the students she leads? This is a vexing question. For instance, does her intellectual and political investment in small-scale farming stifle critique of what non-farming community members, or the students, may see to be dead-end vocations? Or do these other ways of knowing open windows for students on new structures of knowledge and new forms of value? We hope what we offer is a "pedagogy of possibility" (Giroux and Simon 1988), encouraging students to critically engage with forms of knowledge that lie outside their experience and to critically reflect on their own investments – and those of instructors and peers.

Communities are forged through conflict and exclusion as well as consensus and belonging. Working through our community contacts, developed over the course of our research programs, is a process of curation that inevitably entails granting privileged attention to some segments of the community while silencing others. For instance, Elizabeth has two main points of entry to the villages. One is through the women at the farm described above, where she does research on livelihoods, gender, and small-scale farming. The other is through members of the local traditional authority, a "customary" – and contested – political entity comprised of the chief (*hosi*) and her officials. A stay in the villages entails a letter of invitation from the hosi followed by a letter of permission from the police. These permissions are arranged in advance by interpreter Basani Ngobeni – a crucial participant in and co-organizer of the field school.[2] On the field school in Nicaragua, students will mainly observe the practice of what is called *teatro popular*, a form of social theatre taking place in streets, plazas, and open-air festivals, as opposed to mainstream theatre in traditional performance venues. The students will see a mainstream theatre performance and visit a range of theatres, but the purpose of this curated learning experience is to help them learn about the role and significance of applied theatre in the Global South.

These connections have consequences. In the community in South Africa, the traditional authority might be seen as gatekeeper, although it has been so in a positive sense, as facilitator of the field school (Heller et al. 2011). Elizabeth's linkages with these officials have provided a congenial welcome – including colourful formal ceremonies of welcome for the students – and have opened many doors. Students gain ready access to community development projects initiated or supported by the hosi, and knowledge of our connections predisposes other groups to welcome us as well. Yet, the personal nature of these ties is a mark of both resilience and fragility: Elizabeth has built positive relationships with this particular chief and her administration, but what happens when a new hosi takes power? Or when a new field school instructor takes over? The dependence of Elizabeth's research on the favour of the traditional authority makes it awkward for participants to openly critique that body. Even behind closed doors, Elizabeth's status as "expert" and instructor likely curtails students' willingness to do so (Elder 1999; Oberhauser 2002). Finally, her perceived allegiances may have closed other doors in the villages.

Ways to offset these forms of silencing include providing students with a wide overview of relevant research in advance – for example, a complete overview of the theatre industry in Nicaragua, including literature on the mainstream industry, teatro popular, and studies of urban and rural popular economies, as well as critical analyses of traditional or communal power structures in South Africa. In addition, being transparent about our process of curation, its motivations, and the reasons behind certain silences or omissions will provide valuable learning opportunities for both students and those designing and leading the programs.

Destabilization: Radical Difference

Rural women are among the most disadvantaged citizens of South Africa. Marginalized by colonialism, including in its apartheid variant, and more recently by a globalizing economy that seeks to render their food production redundant, they are among the most likely to be impoverished (Mosoetsa 2011). Elizabeth's goals in prioritizing rural women and farmers as participants are several. She aims to expose privileged Canadian students to a way of life shaped by colonial pasts and ongoing global inequities. That intention is signalled by the title of the course. With colonial legacies in mind, we aim, at the outset, for a jarring encounter with difference. Arriving in Cape Town, we spend our first, exhausted day on the ground in a township, an experience one student described as "something that shook us to the core – but something we needed to experience" (Tori, pers. comm.; see also Gmelch and Gmelch 1999; Herrick 2010).[3]

As in many Latin American countries, Nicaraguan teatro popular artists are also marginalized and perceived as second-class citizens: "people look at us as alcoholics, drugs users and governmental aid abusers" (Douglas Mendoza, pers. comm., 11 Feb. 2006). Kirsten prioritizes working with disadvantaged teatro popular practitioners during the field school in order to introduce students to the social complexities of marginalized communities. Teatro popular practitioners, who work as farmers, tourist guides, or journalists, prefer to call themselves "theatre artists" (Ernesto Soto, pers. comm., 22 Jan. 2006) and articulate a desire to talk about the aesthetic values of their practice. They explicitly struggle with their position within the Nicaraguan theatre landscape, and many have expressed confusion about the artistic value of teatro popular, arguing that it deserves to be recognized as legitimate art but with aesthetics distinctive from mainstream art. In informal meetings with these theatre artists, field school students gain insight into the recurring tension between a mainstream discourse of utilitarianism and the desire of those involved in teatro popular to talk about their own aesthetics. In Nicaragua established conceptions of quality in the arts militate against the support of the arts' practices of marginalized communities. Often, traditional criteria of assessment are not seen as appropriate and disallow consideration of popular art or creative works lying outside the mainstream. The Applied Theatre Field School reflexively highlights the social complexities between teatro popular and mainstream arts' practices, introducing students to radically different ways of looking at theatre.

Student reflections on field school learning routinely point to the value of destabilization (Herrick 2010). As one student observed in a reflective essay after the South Africa trip:

> I will never forget how truly bizarre it felt to spend the day picking ground-nuts from the dirt with some of the country's most disadvantaged citizens, and then to eat pizza by an infinity pool that night. (Faelan, pers. comm.)

"Truly bizarre" highlights the dilemma: as curators and leaders of field schools to the Global South, we risk confirming and reifying stereotypes and negative expectations (Glass 2015; Lemmons 2015; Nairn 2005). At the same time, the phrase speaks to the dissonance that can lead students to reflect productively on their position as mobile Northerners merely passing through this space – fleetingly, yet long enough to bear personal witness to the disparities that increasingly define global political economy (Piketty 2014).

Disrupting Difference

Ultimately, the purpose of playing into students' expectations of disorder and dysfunction is to begin their disruption. After seeing a pastiche of some of the most challenging aspects of life on the ground in South Africa – urban and rural poverty, inadequate housing, lack of economic opportunity – the group then spends time with people working to address these challenges. Their efforts and insights help to intervene in the kinds of essentializing discourses visitors may bring with them from the Global North. As student Tori noted:

> The number of community projects and the dedication of the people involved with them debunked the stereotype of the idle, uneducated African unable to help himself, which in turn is used to legitimate the Western "white saviour complex" obsessed with "saving" Africa. (Pers. comm.)

Among essentializing discourses applied to the Global South are those depicting the meanness of what are often called "survivalist" livelihoods (Scoones 2015). In Cape Town, housing activists tell us about the communities that grow up in informal settlements, how community members help hide each other's property when tipped off that the police are coming, and how they play music together in the shelter of night. Students hear more about violence than some can bear – the matter-of-factness with which activists discuss their experiences of rape and gun cultures is unsettling. At the same time, personal stories provide orientation and induce empathy (Mitussis and Sheehan 2013). A brother joined a gang because "there was no one looking after him" (University of Cape Town Workshop 2014): a violent youth becomes a member of a family, a forgotten child, a casualty of lack of social infrastructure.

In the South African villages, young farmers tell us that when they graduated from high school they had no intention of taking up a vocation meant for "the grannies." In informal conversations and interviews – situations in which they are clearly the experts, proud of their knowledge and of visitors' interest in their lives – they reveal that they had hoped to work in offices, be professionals, lead comfortable lives. Starting from a mutual sense of stark difference, South African and Canadian participants enjoy the emerging sense of shared interests. Young people talk about:

> our favourite television shows, favourite pastimes, dreams, what growing up was like ... It cut through cultural differences and allowed us to become true friends. (Kyle, pers. comm.)

Yet, difference is never far from the surface, particularly with respect to access to resources for education and other purposes. Guided by the young farmers' insights, we consider the role of small-scale agriculture in household economies in the South; the structural and personal challenges of rural life in an age of global capital; and barriers to education, urbanization, and other opportunities. By reflecting on shared ground as well as structural differences, students come to see how large a role contingency plays in the lives of those with few material resources. Some begin to think through how Northern lifestyles (e.g., water waste, fossil fuel use, cheap and overabundant food) may contribute to the scarcity and lack that characterize rural life in the South (most immediately, climate change-induced drought and displacement of local foods by cheap, processed imports). These are big steps. They are also tall orders in a short-term field course.

Challenges of Scratching the Surface

Patel (2015) warns of the "shallowness" of understandings drawn from short-term or one-off interactions with distant places and people. We worry about how the structure of a short-term field school may predispose students to draw conclusions based on passing observations. For instance, in South Africa our interactions risk confirming popular preconceptions of African men as indolent, negligent, or absent (Nairn 2005). On the field school we spend more time with women than with men, a bias that is an artefact of Elizabeth's research interests. Students routinely ask questions like "why don't the men help support their families?" When students walk across the village between farm and homestay, they pass two bars where men (and a few women) can be seen in various stages of inebriation at any hour of the day. The many men away at work in urban areas, continuing the colonial-era practice of long-distance labour migration, are out of sight: "productive" men are not much in evidence in the village. Elizabeth makes certain to ask women where their men are and has students read about both labour migrancy and structural unemployment. Nevertheless, comments in reflective essays along the lines of "women are the heart of Africa" and "women hold the continent together" speak to gender stereotypes with a double edge. Hard-working, resilient women are heroic – there is a reason they are such a staple of romantic Africa imagery. But it seems this is a zero-sum game: women's strength connotes male inadequacy.

Underscoring the value of personal interactions in a field school setting, one of the most powerful rejoinders to negative stereotypes about African men is Daniel, the husband of one of the women at the farm. Retired from many years of work at a factory in Johannesburg, Daniel shares personal anecdotes about life as a migrant worker. Daniel's wife Mphephu describes his faithfulness to the family – how he regularly sent back money and was much involved in raising their six children. Daniel and Mphephu are the hosts of our home stay, and Daniel's work repairing things around the women's farm, managing his cattle, and making metal door and window frames (his small business) is much in evidence. Seeing students invoke stereotypes about negligent African men prompts Elizabeth in future to include more "Daniels" among the people we spend time with in the village. Daniel and Mphephu's family story is appealingly familiar to visitors from the North. That very familiarity, though, raises a dilemma. It risks shoring up hegemonic – yet culturally and historically inappropriate – expectations about the nature of family.

The space between difference and sameness is small. We want students to understand that people half a world away, in very different social circumstances, share common aspirations, a common humanity. We hope this knowledge, with the personal valence provided by field school encounters, will move students to take action for social justice. A danger in brief encounters, though, is that visitors grab onto sameness in an uncritical way, much as we seize on recognizable words when hearing a language we don't understand. Looking for shared ground, we risk eliding inequality, sanitizing history, silencing the individual (or the communal). Perhaps the most effective remedy to this tendency is to make it visible. "I wish they would vote but it's not all about me," Liah said in a group discussion after a day at a youth project (pers. comm.). It was the day before the South African national election marking the twentieth anniversary of democracy. Canadian students were puzzled that a substantial number of South Africans of their age were choosing not to vote. We had participated in a debate where the "don't vote" option was articulated powerfully. The observation "it's not all about me" signalled Liah's recognition that there can be valid political positions very distinct from her own. "It's not all about me" is a useful reminder of the constant need to consider how our own experiences and identities shape our readings of others. It is also a reminder of the significance of mediating context, even in encounters that bring us onto common ground. Liah, a politically active student, found it exciting to talk politics with South African students. Before the visit, she and many fellow students anticipated a higher degree of activism among South African youth than among "apathetic" Canadian youth. Yet, she came to recognize that their historical, spatial, and class identities – their difference – might lead rural South African youth to distinctive conclusions about the utility of voting.

Action or Angst?

Another set of questions is raised by our focus on rural and peri-urban locales. By spending most of our time in these predominantly poor communities, what hegemonic narratives of "Africa" and "Central America" are confirmed? By querying the "Africa rising" and "Nicaragua growing" narratives of recent years, repeatedly asking ourselves whether we can see the benefits of GDP growth on the ground, do we puncture youthful optimism about these continents' futures? Yet, by emphasizing the dynamism of diverse rural livelihoods, do we create the impression that resilience is boundless and people can get by just fine without resources? (The pernicious image of the "poor but happy native" comes to mind.) By focusing on inspiring individuals and groups, do we draw a veil over the structures of inequity and injustice that underpin individual and community struggles? Perhaps most damagingly, by examining the shortfalls of both the neoliberal and the pro-poor policy directions of the South African and Nicaraguan governments and civil society organizations, do we leave students who have an interest in development work or other social action in a state of paralysis (Brookfield 2005)? Here what Newstead (2009) calls a "care-full pedagogy" is needed: we must pay close attention to the ways our own political commitments, social locations, and institutional responsibilities impact our students' learning. Instructors' political investments are likely to be more apparent in community settings than they are in the classroom back home. In this context, we need to consider how we can help students develop critical sensibilities in ways that "multiply rather than narrow understandings, open rather than close ways of relating, and enable rather than dis-able modes of engaging" (81).

The Power of Giving Back

Modelling thoughtful, positive engagement with our host communities is one way we attempt to build "critical optimism" (Owens, pers. comm., 19 Aug. 2016). Aware of the often extractive nature of community-based research and concerned that "good intentions" may mask power imbalances inherent in the very structure of short-term study visits (Heller et al. 2011; La Salle 2010; Menzies and Butler 2011; Sidaway 1992), we attempt to compensate the communities in a number of ways. As much as possible, we design field school activities in consultation with host organizations, seeking to include community members as meaningful participants rather than objects of study. For instance, interaction with students in a South African youth program is organized by youth program facilitators as a day of shared classroom and social activity. Days at the farms entail a varied program of farmers teaching students, group interviews, informal conversation, and shared food. In Nicaragua, theatre students volunteer at different cultural organizations where they provide children with drama, dance, and music workshops. They assist in organizing, developing, and constructing new performance venues in La Chispa. We are aware that this approach may fall short of the community-generated activity that is the gold standard of community-based research – activity driven by community interests rather than by the agendas or motivations of outsiders (Carlson, Lutz, and Schaepe 2009; Menzies and Butler 2011). The purposes of short-term field schools may make such community-driven methods elusive. At the very least, activities on our field schools are designed with mutual learning opportunities in mind.

Appropriate material compensation is an essential aspect of ethical research practice in resource-poor communities. No matter how respectful and inclusive our activities, we are asking artists, farmers, and people working in the informal economy to take time out of paid or household work to serve our needs. We do not find it sufficient to give gifts from home, well received as these generally are. Money is what people need. Cash donations to the community organizations that welcome students as volunteers are built into the program fee for both our field schools. Students are encouraged to fundraise in advance for a portion of these monies. They make donations as a group, at levels they determine, to local agencies and grassroots projects including a job-readiness program for youth, theatre troupes, food-security projects, crafts collectives, and schools. Fundraising exposes students to the challenges of soliciting support for grassroots organizations (as opposed to high-profile global agencies like UNICEF or World Vision). At its best, fundraising requires students to consider the differences between human rights and capacity-building approaches to development *versus* charity (Scoones 2015). While there is certainly a tokenism to the volunteer work – how much use to a vegetable project is a student who has never seen a hoe? – financial donations have clear benefit.

At the same time, requiring students to fundraise can lay bare disparities and create inequities within the group. Students of lower income, who may be working at two or three jobs to pay for the trip, have little time to devote to fundraising. Nor can they easily tap family for donations. Among other activities, the 2014 South Africa field school ran a three-month bottle drive across our city, something to which everyone could contribute. At least one student remarked that bottle deposits paid locally might better be ploughed back into local initiatives in our home city. This response speaks to Robson's observation that "sometimes students with their youthful idealism are our best critics in challenging the ethical aspects of arranging field courses to developing countries" (Robson 2002, 336).

In many instances, the most significant contribution of field schools to host communities is employment creation. This is certainly a benefit in rural South Africa and Nicaragua, areas plagued by widespread joblessness. In South Africa, we hire the interpreter discussed above, a driver, and security guards, and we pay the host family's adult daughter for the food (locally sourced) and hospitality she arranges. She considers it culturally inappropriate for us to pay her outright – "You cannot pay, you are our guests" – so in 2014 appreciative students directed some of their raised funds to help her upgrade the sewing machine for her small business. In Nicaragua, we pay local artists, drivers, and host families for the hospitality they provide us in La Chispa.

The downside of such a short-term flurry of activity is highlighted by our difficulty finding a local driver in South Africa. These villages are far off the tourist track and offer little infrastructure for visitors. Drivers with a van and public transport licence are already occupied transporting local people to work and school; appropriately, they are unwilling to abandon long-term clients for a short-term contract, no matter how well it might pay. Such logistical hurdles remind us that determined as we may be to ensure local input and benefit, the fleeting presence of fifteen students and accompanying faculty is a mixed blessing. The cost side of the ledger is front of mind as we write in 2016. Much of South Africa is emerging (one hopes) from the worst drought in generations. Nicaragua is in the midst of an outbreak of the Zika virus. Arriving with an entourage of students this year in Limpopo Province or La Chispa may not have been ethical – or even practical. Instructors need to be responsive to local conditions and maintain the flexibility to cancel field schools to troubled places, the range of which seems ever widening.

Conclusion

Instructors leading field schools to the Global South and other colonized spaces need to take special notice of how their activities are "placed" – South in relation to North, marginalized communities in relation to privileged short-term visitors. Despite the kinds of challenges discussed here, in the end we are convinced, and reassured, by community hosts who exclaim that they look forward to the next visit; and by students who say their experiences in these communities "allowed [them] to purposely set aside [their] world view, however briefly" (Patrick, pers. comm.) or "gave [them] a clearer perspective on how colonialism operates in [their] own country" (Laura, pers. comm.). It seems well worth the effort of confronting these ethical and practical dilemmas if our field schools can contribute to the development of empathetic and critically informed global citizens – people with a sense of expansive "moral and ethical geographies" (Robson 2002) and an understanding that their actions at home and abroad have consequences.

In considering the dilemmas faced by faculty members in relation to host communities and our students, we do not pretend to have an unobstructed view of the terrain. Like our students, we bring to these settings questions and preoccupations that reflect our subjectivities. We do our best to consider the aims and interests of host communities, but we have access to these through the well-worn routes of our own research agendas, and through the skewed power relations that grant us extraordinary mobility when our hosts have so little. We attend to the fissures and exclusions that define and sometimes unsettle communities, but we have our favourite community members, and we know only what we are permitted to see. In the end, the tensions remain pronounced. The best we can do is to attend generously and openly to the voices of our hosts and of the students with whom we take this journey of possibility.

NOTES

The authors lead field schools to South Africa and Nicaragua, respectively. Although the Nicaragua field school has been extensively planned across several years, the course has been delayed because of the Zika virus outbreak. Discussion of planning dilemmas is drawn from both field schools, while in situ examples necessarily focus on the South Africa field school.

1 "Global South" has become the most widely used descriptive label for that category of countries, most of which are former colonies of European powers and many of which are located to the south of those powers. Countries of the Global South are home to the majority of the world's population and, partly as a result of colonial relationships, are marked by high levels

of poverty as measured by gross national income, gross national product, the Human Development Index, and other indices (Schafer, Haslam, and Beaudet 2012; Ravaillon 2011). The modifier "Global" clarifies that the category is not primarily geographical, but defined also in terms of global structural relations of poverty, wealth, and resource access. The label is not uncontroversial, but we accept it for its relevance to South Africa and Nicaragua and in preference to terms like "Third World" and "developing world," with their implications, respectively, of marginality and progress towards some Western/Global North ideal.

2 The field school's reliance on her skills and diplomacy cannot be overstated. More generally, the reliance of community-based field schools on particular personalities bears noting.

3 Student comments are drawn from the post-travel reflective essays of participants in the 2014 South Africa field school. They are reported here as personal communications.

REFERENCES

Alexander, Jacqui M., and Chandra Talpade Mohanty, eds. 1997. *Feminist Genealogies, Colonial Legacies, Democratic Futures*. New York: Routledge.

Anderson, Benedict. 1983. *Imagined Communities: Reflections on the Origin and Spread of Nationalism*. London: Verso.

Appadurai, Arjun. 2000. "Grassroots Globalization and the Research Imagination." *Public Culture* 12 (1): 1–19. https://doi.org/10.1215/08992363-12-1-1.

Arnold, David. 2000. "'Illusory Riches': Representations of the Tropical World, 1840–1950." *Singapore Journal of Tropical Geography* 21 (1): 6–18. https://doi.org/10.1111/1467-9493.00060.

Bakhtin, Mikhail M. (1975). *The Dialogic Imagination: Four Essays*, edited by Michael Holquist. Austin: University of Texas Press.

Bauman, Zygmunt. 2000. *Community: Seeking Safety in an Insecure World*. New York: Wiley.

Bhabha, Homi K. 1994. *The Location of Culture*. London: Routledge.

Booth Fowler, Robert. 1991. *The Dance with Community: The Contemporary Debate in American Political Thought*. Lawrence: University Press of Kansas.

Broekman, Kirsten. 2014. "The Meaning of Aesthetics within the Field of Applied Theatre in Development Settings." PhD diss., University of Manchester, Manchester, UK.

Brookfield, Stephen. 2005. *The Power of Critical Theory: Liberating Adult Learning and Teaching*. San Francisco: Jossey-Bass.

Carlson, Keith, John Lutz, and David Schaepe. 2009. "Turning the Page: Ethnohistory from a New Generation." *University of the Fraser Valley Research Review* 2 (2): 1–8.

Clifford, James. 2013. *Returns: Becoming Indigenous in the Twenty-First Century.* Cambridge, MA: Harvard University Press. https://doi.org/10.4159/9780674726222.

Davidoff, Leonore, Megan Doolittle, Janet Fink, and Katherine Holden. 1999. *The Family Story: Blood, Contract and Intimacy 1830–1960.* London: Longman.

Elder, Glen. 1999. "'Queerying' Boundaries in the Geography Classroom." *Journal of Geography in Higher Education* 23 (1): 86–93. https://doi.org/10.1080/03098269985632.

Evans, Peter B., and William H. Sewell. 2013. "Neoliberalism: Policy Regimes, International Regimes, and Social Effects." In *Social Resilience in the Neoliberal Era,* ed. Peter A. Hall and Michèle Lamont, 35–68. Cambridge: Cambridge University Press. https://doi.org/10.1017/CBO9781139542425.005.

Federal Research Division. 1999. *History of Nicaragua.* The Country Studies Series. Washington, DC: Library of Congress.

Garver, Eugene. 1978. "Rhetoric and Essentially Contested Concepts." *Philosophy & Rhetoric* 11 (3): 156–72.

Geertz, Clifford. 1983. *Local Knowledge: Further Essays in Interpretive Anthropology.* New York: Basic Books.

Giddens, Anthony. 1984. *The Constitution of Society: Outline of the Theory of Structuration.* Cambridge: Polity Press.

Giroux, Henry A., and Roger I. Simon. 1988. "Schooling, Popular Culture, and a Pedagogy of Possibility." *Journal of Education* 170 (1): 9–26. https://doi.org/10.1177/002205748817000103.

Glass, Michael. 2015. "Teaching Critical Reflexivity in Short-term International Field Courses: Practices and Problems." *Journal of Geography in Higher Education* 39 (4): 554–67. https://doi.org/10.1080/03098265.2015.1084610.

Gmelch, George, and Sharon Bohn Gmelch. 1999. "An Ethnographic Field School: What Students Do and Learn." *Anthropology & Education Quarterly* 30 (2): 220–7. https://doi.org/10.1525/aeq.1999.30.2.220.

Hall, Peter A., and Michèle Lamont, eds. 2013. *Social Resilience in the Neoliberal Era.* Cambridge: Cambridge University Press. https://doi.org/10.1017/CBO9781139542425.

Hall, Stuart. 1997. "The Local and the Global: Globalization and Ethnicity." In *Culture, Globalization, and the World System,* ed. Anthony D. King, 19–39. Minneapolis: University of Minnesota Press.

Harries, Patrick. 1989. "Exclusion, Classification and Internal Colonialism: The Emergence of Ethnicity Among the Tsonga." In *The Creation of Tribalism in Southern Africa,* ed. Leroy Vail, 82–117. London: James Currey.

Hay, Michelle. 2014. "A Tangled Past: Land Settlement, Removals, and Restitution in Letaba District, 1900–2013." *Journal of Southern African Studies* 40 (4): 745–60. https://doi.org/10.1080/03057070.2014.931062.

Heller, Elizabeth, Julia Christensen, Lindsay Long, Catrina A. Mackenzie, Philip M. Osano, Britta Ricker, Emily Kagan, and Sarha Turner. 2011. "Dear Diary: Early Career Geographers Collectively Reflect on Their Qualitative Field Research Experiences." *Journal of Geography in Higher Education* 35 (1): 67–83. https://doi.org/10.1080/03098265.2010.486853.

Herrick, Clare. 2010. "Lost in the Field: Ensuring Student Learning in the 'Threatened' Geography Fieldtrip." *Area* 42 (1): 108–16. https://doi.org/10.1111/j.1475-4762.2009.00892.x.

Hoggett, Paul. 1997. "Introduction: Contested Communities." In *Contested Communities: Experiences, Struggles, Policies*, ed. Paul Hoggett, 1–22. Bristol: Policy Press.

hooks, bell. 1989. *Talking Back: Thinking Feminist, Thinking Black.* Boston: South End Press.

hooks, bell. 1997. *Wounds of Passion: A Writing Life.* New York: Henry Holt.

Kepe, Thembela. 1999. "The Problem of Defining 'Community': Challenges for the Land Reform Program in Rural South Africa." *Development Southern Africa* 16 (3): 415–33. https://doi.org/10.1080/03768359908440089.

Kymlicka, Will. 1995. *Multicultural Citizenship: A Liberal Theory of Minority Rights.* New York: Oxford University Press.

La Salle, Marina J. 2010. "Community Collaboration and Other Good Intentions." *Archaeologies* 6 (3): 401–22. https://doi.org/10.1007/s11759-010-9150-8.

Lemmons, Kelly. 2015. "Short-Term Study Abroad: Culture and the Path of Least Resistance." *Journal of Geography in Higher Education* 39 (4): 543–53. https://doi.org/10.1080/03098265.2015.1084607.

Li, Tania M. 1996. "Images of Community: Discourse and Strategy in Property Relations." *Development and Change* 27 (3): 501–27. https://doi.org/10.1111/j.1467-7660.1996.tb00601.x.

Lugones, María. 2007. "Heterosexualism and the Colonial/Modern Gender System." *Hypatia* 22 (1): 186–219.

McMorran, Chris. 2015. "Between Fan Pilgrimage and Dark Tourism: Competing Agendas in Overseas Field Learning." *Journal of Geography in Higher Education* 39 (4): 568–83. https://doi.org/10.1080/03098265.2015.1084495.

Menzies, Charles R., and Caroline F. Butler. 2011. "Collaborative Service Learning and Anthropology with Gitxaała Nation." *Collaborative Anthropologies* 4 (1): 169–242. https://doi.org/10.1353/cla.2011.0014.

Mignolo, Walter D. 2000. *Local Histories/Global Designs: Coloniality, Subaltern Knowledges, and Border Thinking.* Princeton, NJ: Princeton University Press.

Mitussis, Darryn, and Jackie Sheehan. 2013. "Reflections on the Pedagogy
    of International Field-schools: Experiential Learning and Emotional
    Engagement." *Enhancing Learning in the Social Sciences* 5 (3): 41–54. https://
    doi.org/10.11120/elss.2013.00013.
Mosoetsa, Sarah. 2011. *Eating From One Pot: The Dynamics of Survival in Poor South
    African Households.* Johannesburg: Witwatersrand University Press.
Nairn, Karen. 2005. "The Problems of Utilizing 'Direct Experience' in Geography
    Education." *Journal of Geography in Higher Education* 29 (2): 293–309. https://
    doi.org/10.1080/03098260500130635.
Newstead, Clare. 2009. "Pedagogy, Post-coloniality, and Care-full Encounters
    in the Classroom." *Geoforum* 40 (1): 80–90. https://doi.org/10.1016/
    j.geoforum.2008.04.003.
Oberhauser, Ann. 2002. "Examining Gender and Community through Critical
    Pedagogy." *Journal of Geography in Higher Education* 26 (1): 19–31. https://doi
    .org/10.1080/03098260120110340.
Patel, Kamna. 2015. "Teaching and Learning in the Tropics: An Epistemic
    Exploration of 'The Field' in a Development Studies Field Trip." *Journal
    of Geography in Higher Education* 39 (4): 584–94. https://doi.org/10.1080/
    03098265.2015.1084499.
Piketty, Thomas. 2014. *Capital in the Twenty-First Century.* Cambridge, MA:
    Harvard University Press. https://doi.org/10.4159/9780674369542.
Ravaillon, Martin. 2011. *Global Poverty Measurement: Current Practices and Future
    Challenges.* Washington, DC: Development Research Group of the World
    Bank.
Robson, E. 2002. "An Unbelievable Academic and Personal Experience:
    Issues around Teaching Undergraduate Field Courses in Africa." *Journal of
    Geography in Higher Education* 26 (3): 327–44. https://doi.org/10.1080/
    03098260220000019909.
Said, Edward W. 1993 [1978]. *Culture and Imperialism.* New York: Vintage.
Said, Edward W. 2003. *Orientalism.* Rev. ed. New York: Vintage.
Schafer, Jessica, Paul Haslam, and Pierre Beaudet. 2012. "Meaning, Measurement,
    and Morality in International Development." In *Introduction to International
    Development,* ed. Paul Haslam, Jessica Schafer, and Pierre Beaudet, 3–26. Don
    Mills, ON: Oxford University Press.
Scoones, Ian. 2015. *Sustainable Livelihoods and Rural Development.* Halifax: Fernwood.
    https://doi.org/10.3362/9781780448749.
Sidaway, James Derrick. 1992. "In Other Worlds: On the Politics of Research by
    'First World' Geographers in the 'Third World.'" *Area* 24 (4): 403–8.
Spivak, Gayatri Chakravorty. 1988. "Can the Subaltern Speak?" In *Marxism
    and the Interpretation of Culture,* ed. Cary Nelson and Lawrence Grossberg,

271–313. Urbana: University of Illinois Press. https://doi.org/10.1007/978-1-349-19059-1_20.

University of Cape Town Workshop on Urban Housing Issues for Field School Students. 2014. African Centre for Cities, University of Cape Town. 2 May.

Urry, John. 2002. *The Tourist Gaze.* 2nd ed. London: Sage.

Vibert, Elizabeth. 2016. "Gender, Resilience and Resistance: South Africa's Hleketani Community Garden." *Journal of Contemporary African Studies* 34 (2): 252–67. https://doi.org/10.1080/02589001.2016.1202508.

Westlake, E.J. 2005. *Our Land Is Made of Courage and Glory: Nationalist Performance of Nicaragua and Guatemala.* Carbondale: Southern Illinois University Press.

Wheeler, Roxann. 2000. *The Complexion of Race: Categories of Difference in Eighteenth-Century British Culture.* Philadelphia: University of Pennsylvania Press. https://doi.org/10.9783/9780812200140.

WHAT YOU CAN'T GET FROM A TEXTBOOK ...

The law is innately black and white; it requires lawyers to exist in a state of impartiality, weighing only the logical thought process of the court, not the people behind each decision. As law students, we learn in black and white. We begin to absorb those ideals and make them a part of us. We observe the plaintiffs and defendants as if they are characters in a play while we remain in the ivory tower, looking down on a landscape of monochrome.

Before arriving in Bella Bella, home of the Heiltsuk First Nation, all I knew about the town was what the law taught me. In the case of *R. v. Gladstone*, a groundbreaking decision in Aboriginal law, the court affirmed the Heiltsuk right to commercially harvest herring roe. Objectively, I understood the application of various legal tests, the factual basis behind the court ruling, and the rationale of the judge. This is how you learn as a law student. This is what you learn in law school; for most, this is all we ever know.

Upon arrival in Bella Bella, I was able to see what didn't make it to the textbook pages. The Heiltsuk way of life revolves around the sea. I already knew the sale, trade, and barter of fish for commercial purposes was an integral part of Heiltsuk culture. But that's only what Justice Lamer wrote in our legal black and white in *R. v. Gladstone*. What he didn't see was how along the streets of Bella Bella the homes face towards the ocean, the people flock along the shore, and even the flags proclaiming "No Enbridge" seem to shout towards the sea. As we wandered through the town, a young boy ran across the road, whipping the tail of a bull-kelp. He paused, looked in our direction, and blushed, darting off as quickly as he had come, his bull-kelp following him in the breeze as he ran. This is the very kelp that his ancestors would paddle to, harvesting the herring spawn speckled along the blade. This kelp has supported their community since time immemorial, providing food and tradeable goods. This is the foundation of the Gladstone decision affirming commercial harvesting rights to this First Nation. This green blur never made it to the pages of the textbook, but it is what truly mattered to those the case affected.

A field course in the context of a legal education is a chance to learn more than the black and white of a statute or case decision. It provides the colour that gets bleached from the pages of our textbooks and brings life back to the law. It's a chance to see the bull-kelp, to observe first-hand the significance of the herring roe, and understand the implication of a ruling from a perspective other than that of our text-book pages.

Sarah Elwood
Participant, University of Victoria's Hakai Field
Course in Environmental Law and Sustainability

SECTION THREE

Assessing the Value of the Journey

Although some form of evaluation is implied in many chapters in this volume, this section explicitly considers the process of assessing impacts and outcomes of learning out there. The three chapters, by authors from different disciplines and universities, start from a similar place but stake out distinctive terrain. Each author admits that assessing the value of a field study presents a daunting challenge, but insists that such an undertaking is critical to ensuring such programs deliver on their promise and receive continued support. The chapters remind us that we cannot simply assume our teaching and learning in the field is having a positive impact. Each chapter offers helpful guidance to those tasked with undertaking assessment and, in each case, authors connect their reflections to a specific analysis of their own field programs.

Chapter 8, "Getting Beyond 'It Changed My Life,'" is by clinical psychologist Janelle Peifer and Elaine Meyer-Lee, a developmental psychologist. The two hail from Agnes Scott College, a private, liberal arts women's college in metropolitan Atlanta, Georgia. The chapter provides a useful summary of considerations for developing a "thoughtful, targeted process" for assessing the transformation that may occur for students during study abroad. Peifer and Meyer-Lee organize the first part of their chapter around four types of transformation – academic, socioemotional, host community, and institutional – each with a useful set of questions that could be employed by would-be assessors. They follow this typology with a four-step guide to undertaking an assessment and ground the chapter in a case study detailing the design and assessment of their own Journeys travel-study program.

Chapter 9, by Cameron Owens and Maral Sotoudehnia, geographers at the University of Victoria, is also focused on practical guidance with a critical eye on the complex challenges of assessing the impacts of "out

there" learning. The chapter is organized around four specific challenges that evaluators often face: marshalling the time and resources
to undertake comprehensive evaluation, determining what elements to
consider, producing reliable data, and undertaking qualitative analysis
in the context of uncertainty and political complexity. Drawing on their
experience delivering and assessing urban sustainability field schools in
North America and Europe, the authors point to some creative possibilities for addressing such challenges.

The final chapter in this section, by Michael Glass, a geographer in the
Urban Studies Program at the University of Pittsburgh, is perhaps the most
critical – or healthily sceptical – of the three. While the other two chapters
are not naively uncritical about the challenges of assessment, Glass queries the very notion of "transformation." The chapter is organized around
three tropes – transformation as brand, transformation as outcome, and
transformation as theory – calling into question (although not dismissing)
transformation as the ostensible goal of field learning. Glass invites us to
inquire into the political work that "transformation" does in the current
neoliberal context of internationalizing the university and to attend to
how transformative learning has been theorized. Like the other chapters
in this section, this one provides a useful overview of a specific field course,
Glass's Urban Studies Field School to cities in Southeast Asia.

While every author in this section (and indeed in this volume) sees
great promise in travel study programs, each reminds us that we must
keep a critical eye and an open mind to the challenges of understanding –
and assessing – the outcomes of learning out there.

Four student vignettes round out this section. They demonstrate the
creative ways participants chose to reflect on their off-campus experiences. Field school student and later teaching assistant Andrea van Noord
recalls a productive moment of destabilization and self-recognition while
staring into a granite pit at a concentration camp in Austria. Jake Noah
Sherman, also part of a field school studying Holocaust memorialization,
uses the genre of poetry to characterize his experiences in Berlin on
14 May 2016. And then in a short creative writing piece, Sherman illustrates
how initial resistance in a personal encounter can turn to openness and
shift one's outlook on identity. Liah Formby reflects on the difficulties of
explaining the complexities she grappled with during her time in South
Africa on a field school studying the legacies of colonialism, and Emily
Tennent describes a realization she reached about education while in
India on an Applied Theatre program. Each of these vignettes portrays a
sense of learning and growth that is both personal and academic.

EYES WIDE OPEN ...

The heat seems nowhere more oppressive than when radiating from the stone structures of a former Nazi concentration camp. Standing at the edge of the infamous granite quarry at Mauthausen near Linz, Austria, palms against the angled glass wall of a memorial, I had the sensation of leaning forward as if to fall into the quarry. As waves of heat and vertigo ran the course of my body, I surveyed the 186 stairs up which prisoners would carry roughly hewn stone blocks, weighing as much as 50 kilograms. Behind me, I heard our tour guide describe a favourite game played by the guards. Made to stand in a line at the edge of a cliff known playfully as the "parachutists wall," prisoners were instructed to push the individual in front of them over the side or be shot. As was often the case, the story invoked no immediate response from our class, so colossal was the image and so horrific was the predicament.

Later, we found ourselves sitting in a misshapen circle, as if it, too, was warped by the heat of our excursion, attempting to "process" our experience. As we went around the room, each voicing our thoughts, trying to characterize our state of mind and respond meaningfully to what had been described to us at Mauthausen, I heard the words: "I would never do that. I would never push the person in front of me. I couldn't, and I would not." I remember my classmate's eyes as the words tumbled out; wild with conviction, desperate to find himself on the right side of an imaginary line and to demonstrate having *learned* something from the story. I remember feeling panicked by his words as if they were dangerous, as if they came from a place of hysteria, not the mouth of a kind, responsible, empathetic young man who only a handful of years later would become a father. I think of this moment often, not because it is exceptional, but because it proved only to be the first expression of what I now understand to be the inevitable and, for some, the central challenge of the field school: to know that we can all be reduced to states and behaviours in which we do not recognize ourselves.

No longer a student, I am now the Experiential Learning Facilitator for the same program. My job, as I understand it, is to help students to learn amid the experience of place, amid a material world that makes real the facts and figures of history, giving them contours and faces *just like ours*. I believe it is my responsibility to try and hold students in the ferocity of such moments, to ask them to slow down, to sustain their discomfort and to understand both the futility and the danger of measuring this particular past against the moral aspirations of a more familiar world – without slipping into the rabbit hole of relativity. It is a fine, and at times impossible, balance to strike, but it is a confrontation with the self from which I have seen students emerge – time and again – with a steadiness and stillness that are the marks of the individual with eyes wide open.

Andrea van Noord
Participant and Facilitator, University of
Victoria's I-witness Holocaust Field School

8

Getting Beyond "It Changed My Life": Assessment of Out There Transformation

JANELLE S. PEIFER AND
ELAINE MEYER-LEE

Off-campus study programs have been heralded for their transformative potential. Over the past decade, short-term global programs have grown more than any other type of undergraduate student study abroad experience, with 63% of students studying abroad now doing so for less than eight weeks (Institute of International Education 2016). Although brief, these experiences can help broaden students' intercultural sensitivity (Anderson et al. 2006), encourage their personal growth (Chieffo and Griffiths 2004), and increase their knowledge and skills (Kehl and Morris 2008). Yet, capturing the nature of transformation is a difficult undertaking. For one thing, the definition of transformation itself is elusive and ambiguous, as Glass so helpfully unpacks in his chapter in this volume. In addition, off-campus study programs vary widely with vastly different objectives, durations, leaders, and student participants. One program may take first-year students to Nicaragua to learn more about the role of globalization in the international coffee trade, while another may study comparative research on identity in Croatia and the United States with fourth-year psychology majors. Certainly, these two programs expect vastly different outcomes and thus require different assessment plans.

A thoughtful, targeted assessment process serves many purposes. It can help articulate the extent to which students transform through their off-campus study programs, concretize target outcomes for participants and stakeholders, support broader institutional goals, and encourage continued improvement and refinement of programs. One essential step in the process of designing assessments is identifying the goals of the programs. This can be done in a more hierarchical way (e.g., the professor outlining the learning objectives for the program in the syllabus and communicating them to students), in a more collaborative manner (such as students brainstorming their personal development and learning goals for the program in small teams), or in a way that combines both approaches. After identifying these objectives, one can select appropriate measurement techniques and tools to capture the outcomes of interest.

These assessment processes can be enacted at varying levels and with different scopes (e.g., an individual instructor assessing their students' learning or an institution-wide comprehensive assessment to better understand the role of intercultural competence in student retention). To capture the nuanced impact of students' transformation, many assessors have moved beyond pre- and post-surveys alone to employ comprehensive, multimethod, ongoing assessment techniques (Deardorff 2015). This approach can help isolate the aspects of students' globally related experiences (including academic, social, and travel-based), personal characteristics (such as personality traits or ethnic identity), and prior experiences that shape outcomes and help institutions better understand how off-campus study programs affect students over time.

Despite the many potential benefits of assessment, many international educators view assessment as one of the most daunting or frustrating parts of off-campus study programs. This chapter, together with the others in this section, aims to help. First, we explore aspects of transformation that often occur in off-campus study programs with a focus on the following four specific areas: (1) academic, (2) socio-emotional, (3) host community, and (4) home institutional transformation. Next, we discuss ways that program leaders can individualize the definition of transformation to be meaningful for their programs. Then, we walk through techniques to assist in developing an assessment plan and identify struggles associated with this process. Finally, we present an illustrative case study of a longitudinal, multimethod assessment plan undertaken at a small liberal arts college for women in the southeastern United States. In the case study, we discuss concrete strategies for international education assessment and lessons learned from this project.

Aspects of Transformation

Developing a universal definition of transformation for off-campus study experiences is untenable. Even the term "transformation" brings to mind the idea of change into a final product stagnated in a fixed state. For the purpose of this chapter, we conceptualize transformation as a vivid, dynamic developmental process that may encapsulate small change, growth – and even at times – regression. This definition is closest to Glass's category of transformation viewed in terms of "learning out-comes" (in this volume) and acknowledges the concerns he raises there.

We would, in fact, argue that education is inherently value-laden, and in this current historical moment of increasing nationalism and isolation-ism, it is more important than ever to make explicit the specific changes or growth we are trying to facilitate through off-campus study programs. With this in mind, the different forms of assessment we discuss here aim to measure and capture at least *some* of this process of change (or lack thereof), especially in service of the stated objectives of the program. We discuss how different assessment methods can serve as tools for exam-ining a wide range of transformation-related aims such as developing students' awareness of intersectionality in their own cultural identity or measuring the cost-effectiveness of a field school program for building language competence.

With this wide range of possible transformations, no single method-ology or definition of transformation will do. Rather, each group must identify their own specific goals related to various factors such as the program, home and host culture, and participants. To identify objec-tives during the program-design stage, trip leaders can consider the ques-tion: "How can this experience transform the way my students think, feel, and act?" Before considering this, program leaders must reflect on how they define transformation for their specific program. This process may be more challenging than it appears at face value. The construct of transformation is elusive, multifaceted, and dynamic. What will it look like for different students and contexts? What unanticipated transforma-tions (both positive and negative) may arise? Responses to questions like these can form the foundation for developing a few specific objectives. This individualized definition of transformation – and the assessable, targeted outcomes that are associated with it – can be developed in sev-eral ways. For instance, outcomes may be shaped by wider institutional strategic goals, or they may be driven by the content and objectives of a specific course. A psychology professor designing a course on mental health accessibility to high poverty communities in rural France will have very different goals for their off-campus study experience than a student affairs professional designing a program on social activism in Beirut. At the same time, a diverse group of programs may share some similar transformative potential, such as increased comfort with difference, self-reflection, or appreciation of the influence of context.

Program leaders can also partner with various stakeholders, such as students, host communities, and their institution's staff and faculty, to identify these transformational objectives. The next step then is to infuse these goals across the course or program experience in meaningful ways. Some research suggests that students who identify specific goals for study abroad experiences are more likely to develop intercultural sensitivity and awareness (Kitsantas 2004; Paige et al. 2006). Students' goal setting as part of the learning process can be informed by factors such as students' background, past international experience, occupational aspirations, and/or their immediate personal aims and desires.

Having introduced some of the challenges of assessing and characterizing "transformation," we turn now to a central concern of our chapter – examining assessment in what we identify as four key realms: (1) academic, (2) socio-emotional, (3) host community, and (4) institutional. We target these realms because they represent four different facets of the field school experience that most students and programs will encounter, regardless of individual differences.

Academic Transformation

Programs that prioritize academic growth can focus their assessment approach on student learning outcomes. This method of assessment focuses on educational goals. Program leaders may choose to focus on students' knowledge about another culture (e.g., its histories and languages) or on culture-general learning (such as broader intercultural competence skills). They may also track students' general academic growth. Will students show higher levels of language competence after participating in a field school experience? Do they have a better grasp of content-specific skills and theories after an applied international experience compared with peers in a traditional classroom? These cognitive skills are an essential part of how students respond to other cultures and students' development as learners and individuals. For example, knowledge about local norms and current events enables students to engage at a deeper level in their cross-cultural settings. Program leaders must carefully select assessment forms that match their overall goals and prioritize which to focus on.

Before choosing learning assessments, the instructor and/or students choose goals that are specific, measurable, and achievable within the context of the course. These student learning outcomes may reflect the individual instructor's course goals, the department's learning objectives, and/or wider institutional learning goals. Once identified, these focused outcomes lend themselves to assessment in myriad ways. We will focus on the use of rubrics to assess academic transformation.

Program leaders can translate their student learning outcomes into rubrics that can be used to assess course assignments (e.g., essays, reflection videos) at different points in the semester and evaluate potential growth and change in this way. Changes in scores in these assignments over time can reveal shifts in student learning, and the course assignments themselves may enable leaders to explore students' process, thinking, and understanding – particularly as they relate to the articulated aims or benchmarks of the course. Using a rubric, evaluators or the students themselves can score selected course requirements. When creating the rubric, program leaders will identify the assessment anchor points and gradients in between. It is important to include a rating scale with descriptive, specific language that captures the standards and expectations for the assignment or activity and allows raters to assess the product on a continuum. Rubrics can also be modified to serve as observational protocols where evaluators observe an action (e.g., a cross-cultural psychological symptom inventory for depression) and rate the students on the rubric scale. See Table 8.1 in the appendix to this chapter for a sample rubric assessing key student learning outcomes for a short-term field school trip.

Some Guiding Questions for Measuring Academic Transformation

1 How discipline-specific and discipline-general do I want students' learning to be in this program?
2 What are three to five things I want students to be able to do, articulate, and/or have a critical awareness of by the end of the program?
3 How will I know if students have succeeded in attaining the learning outcomes? How will I know if they have not?
4 What aspects of the host community destination lend themselves to learning in a way that students could not experience in their home community?

Socio-emotional Transformation

Socio-emotional development appears frequently as a key outcome of off-campus study programs. It refers to students' affective experiences, particularly their emotional expression, and their ability to form meaningful social relationships (Cohen 2006). While traditional courses in students' home institutions may encourage individual growth, study-abroad programs often expose students to experiences that push them to investigate who they are within a broader context. During international experiences, students assess their strengths and weaknesses without the comfort of familiarity. Through these challenges, students often learn to engage with those around them in deeper and more complex ways. As they develop more complex social schemas in novel environments, they investigate their own assumptions and biases, and often, they learn to think more critically about their own place in the wide world. Students who are travelling out of the country for the first time may reflect on their relationship with their home culture in new ways as they encounter a cultural context with different rights and restrictions. For example, a Black American woman travelling outside the US South to New Zealand for the first time may explore new aspects of her sense of self in a culture that lacks her home culture's history of enslavement and segregation of people of African descent.

Capturing and assessing this important form of transformation can be difficult. Similar to academic transformation, trip leaders must first identify the key areas of socio-emotional growth that fit the design and duration of their program and the demographics of their student group. Given the personalized nature of socio-emotional transformation, trip leaders can encourage students to develop or amend goals to match their individual style, personality, and objectives. The success of achieving these goals can then be assessed through methods such as self-report measures (e.g., surveys, mood logging, journalling), observational protocols of live interactions, and/or in-country and post-travel interviews. A mix of objective, qualitative, and quantitative methodologies can illuminate the complex landscape of students' socio-emotional growth. Interviews can be targeted to assess specific outcomes in detail or more open-ended to allow for themes to arise more spontaneously.

For an example of self-report or interview, students may respond to open-ended prompts to capture possible transformation with questions such as:

- Discuss your appreciation and understanding of your own culture and identity.
- Describe a moment during your time abroad that impacted you in a way that was meaningful to you. How did it impact you and why?

Some Guiding Questions for Socio-emotional Transformation

1 What pre-existing intra-individual variables may impact the way that students experience the program (e.g., their socio-economic background, sexual orientation)?
2 What social and/or emotional domains does this program target? What activities or experiences are most likely to shape socio-emotional outcomes?

Host Community Transformation

One of the most overlooked areas of transformation is the change that can occur in the host communities where off-campus study programs take place. These programs can have a lasting impact on the local people and resources – both in positive and negative ways (see Vibert and Sadeghi-Yekta as well as Owens and Sotoudehnia, in this volume, for more in-depth explorations of this issue). Without considering the role and perspectives of the host community, a key element of transformation is neglected. For example, the frequent arrival and departure of student groups can strain the ecological and social balance of communities. A group working with young people who have been orphaned, for instance, must be particularly aware of how brief, inconsistent contact with travellers can interrupt healthy attachment and emotional well-being of children.

Students, faculty leaders, and host community members who work collaboratively can identify tenable, sustainable goals for the off-campus study program. Together, they can develop the most effective methods for assessing progress and measuring both short- and long-term outcomes. By incorporating community stakeholders into the assessment process, program leaders can obtain more thorough data and measure the intended and unintended results of their programs. This more reciprocal, ongoing assessment process allows all parties involved to reflect on and refine their approach to programs in dynamic ways. Figure 8.1 represents a sample host/visitor assessment sheet. Something like this assessment sheet could be used to facilitate collaborative planning and debriefing of field school experiences. Of course, this could be modified significantly to meet the needs and communication style of the host community and visitors.

Complete this worksheet with representatives from the host and visiting communities. First, identify specific, achievable goals that are meaningful to each community. Then, use the rating scale (1: not achieved at all – 7: exceeded expectations) to rate the program's success at achieving the goal for each area.

Groups can work together to achieve consensus on ratings or have representatives complete separate worksheets.

Goal Area	Hosts	Rating	Visitors	Rating
Learning	Host community participants learn 2 alternate methods for water filtration to meet the needs of community centre.	Pre: Mid: Post:	Visitors learn activities and management techniques for running a community centre in Kimende.	Pre: Mid: Post:
Community Action				

Figure 8.1. Sample worksheet for assessing impacts in host community.

Guiding Questions for Measuring Host Community Transformation

1 Who are key members of the host community who can participate in assessment planning? What are their roles and connections to the community?
2 Why have I chosen to collaborate with these particular host community people and excluded other potential participants? What are the costs and benefits of these choices?
3 What potential long- and short-term impacts (both negative and positive) should we anticipate and seek to assess?
4 What assessment method best meets the needs of the host community?
5 How will we share assessment findings with the community? How can we include community feedback in the process of refining our programs over time?

Institutional Transformation

Assessing macro-level change at the institutional level is also key to understanding the impact of off-campus study programs. Individual students are nested within their larger college and university environment.

Institutional variables such as the programs offered, student body demographics, or the university's mission can shape and be shaped by off-campus study experience in a reciprocal way. From a practical standpoint, several institutional units may be invested in capturing how these experiences contribute to wider strategic goals (e.g., internationalization, increasing student retention). Administrators who feel invested in capturing these outcomes are more likely to provide resources to support assessment goals.

Moreover, faculty, trip leaders, and researchers also have an opportunity to gain a breadth of insight beyond their individual trip or classroom. While there are several potential advantages, program leaders must be aware that measuring institutional change is enormously difficult. Change that occurs may be attributed to various concurrent factors that may or may not be measurable. Furthermore, assessment data and findings can generate information that institutional constituents may not like. Those doing assessment may feel pressured to support a certain narrative with their work or to stifle findings that do not align with this narrative.

Program directors completing assessment must be aware of these potential pitfalls and work to establish plans that can help to overcome them. Each institution is different, but program directors may benefit from working as part of an interdisciplinary, diverse team to plan assessment strategies and think through the analyses and framing of findings. Using a multimethod, rigorously developed research design (e.g., including longitudinal data) will help bolster assessors' confidence in the integrity of their findings. Moreover, it can be helpful to remember that unexpected – even seemingly "negative" findings – can serve the purpose of refining and improving programs in the long run. It can be tempting to use data to present and support institutional, individual, or instructional notions and priorities. Yet, one of the major reasons to pursue comprehensive assessment is to uncover potential pitfalls or growth areas for students, instructors, institutions, host communities, and programs as a whole.

Guiding Questions for Measuring Institutional Transformation

1 Does my program align with any wider institutional strategic goals? If so, how?
2 What aspects of the institution (e.g., size, religious affiliation, region) may inform the way students experience my program? How do

institutional factors impact my assessment findings and how do my
findings impact the institution?

3 What institutional stakeholders may be invested in my program and
   its outcomes? For what purposes?

The Process of Assessment

We have been arguing that, while many students return with exclama-
tions about the life-changing nature of their off-campus programs, doc-
umenting evidence of specific change is necessary to understand student growth, inform continuous quality improvement of our pedagogy
and curriculum, and advocate for this type of experiential education.
Research has established that we cannot assume that transformation or
even change will automatically occur as a result of study abroad (Vande
Berg, Paige, and Lou 2012). To truly measure our success, we need to
go beyond touting our efforts and participation levels to capturing outcomes or impact for students and other stakeholders. Students are much
more likely to learn things they have been primed to pay attention to in
advance (Turkay 2014; Moeller, Theiler, and Wu 2012). The simple act
of setting goals itself is correlated with increased success in many areas,
but these goals can also be a useful predictor variable for assessment, as
will be discussed more below. Similarly, having students come back and
reflect on (or self-rate change towards) their own goals after re-entry
can serve as both a reinforcer of change and an assessment measure,
among other things. To summarize, the ideal cyclical process of assessment involves the following steps:

1 **Define outcomes (based on your mission, strategic plans, and goals)
   and establish measurable criteria.**

As mentioned above, there are many different potential aims of off-
campus study programs, and the clearer we are about our own, the better
we can design a program to achieve them. Are we focusing on language
learning? Disciplinary knowledge? Intercultural competence (and if so,
which relevant knowledge, attitudes, or skills)? Other personal growth?
This brainstorming process should include as many of the key faculty
program leaders and/or administrators involved with the off-campus
study program as possible. Once pedagogical goals have been articulated as specific student outcomes, we also need to spell out primary
goals (and audiences) for our assessment itself. For example, depending

on our questions, we might want to look comparatively at various possible factors in differential impact such as program location, model, and duration; student demographic characteristics; or students' own choices about intercultural engagement. If longer-term developmental impacts are a key interest, we may want to pursue a longitudinal approach. Because transformation is a complex process, employing multiple methods is preferable, but it is also essential to inventory our resources realistically and not create an assessment plan so overwhelming that it is not executed.

2 Identify appropriate assessment methods.

Only once outcomes and evidence of success have been defined can we identify appropriate assessment methods. It is tempting sometimes to start with a popular instrument, but this is by no means a one-size-fits-all kind of question. As the old adage says, "If your only tool is a hammer, everything looks like a nail!" In selecting methods, there are many to choose from, but it is important to remember the difference between indirect methods where students report their own changes such as through surveys or interviews (or the self-reflection on goals described above) and direct methods where we are observing changes ourselves through testing, portfolios, observation, and so forth. Either of these types can be quantitative or qualitative in nature. Examples of methods are described in the case study below.

3 Collect data before, during, and/or after the off-campus program.

Of course, the specifics of this phase will depend entirely on the assessment methods chosen in the previous step; again, examples are described below.

4 Analyse data and reflect on needed changes, then design and apply changes (including revising outcomes to start the whole cycle over again).

Our efforts are wasted if we do not actually use the results of our assessment to improve our designs in an ongoing iterative process. For our own study abroad programs, we have seen how our own collected assessment data have helped us choose, advocate, and implement pedagogical and curricular program design elements (incorporating more immersion);

policy changes (such as GPA requirements); advising approaches (taking into account program features and students' goals, ages, ethnicity, etc.); and pre-departure preparation (talking more about goals and engagement choices).

Assessment in Action: A Case Study[1]

Agnes Scott College, a women's college in the Southeastern United States that is highly diverse in racial and socio-economic terms, has incorporated a comprehensive focus on global learning and leadership through a unique college experience called Summit. Every student, regardless of major, completes a core curriculum and co-curriculum focused on global learning and leadership development,[2] develops a digital portfolio, and builds a personal board of advisers to help guide their academic and professional development. The goal of Summit is to prepare every Agnes Scott graduate to be an effective change agent in an increasingly globalized world – or, from our mission statement "to think deeply, live honorably, and engage the social and intellectual challenges of her time."

This case study will focus on one key component of Summit, a required first-year Global Learning course and study tour called Journeys, which builds on an introductory fall semester course. This spring semester course includes an approximately eight-day study tour to locations including Jamaica, Panama, Puerto Rico, Croatia, Northern Ireland, Morocco, Central Europe, England, Martinique, Dominican Republic, Trinidad, Cuba, Bolivia, Nicaragua, Canada, the Navajo Nation, and New York City. Each section of typically seventeen to twenty-one students focuses on a different global destination from this list and on a content topic (economics, public health, women's studies, etc.). The course focuses on building students' intellectual understanding, affective awareness, and intercultural skills. Across all sections of the course, students focus on the following four key global themes: (1) Identity: Self, Culture, and Other; (2) Journeys (ethics of travel); (3) Globalizations; and (4) Colonialism, Imperialism, and Diaspora. These thematic elements drive the pedagogical content of the course and establish a cognitive foundation for intercultural competence with all students. Courses explore these themes through intentionally active, participatory strategies including digital storytelling, community-based projects, and plenary panels and dialogues aimed at translating cognitive awareness into action. The embedded travel component of the course (led by the faculty leader and a co-leader) enables all first-year students to apply and refine more abstract skills in a lived, personalized way. Moreover, students connect wider concepts to a specific location and cultural context.

The course focuses on the dynamic process of knowledge development, namely, developing a better understanding of one's home country. This includes exploring one's home culture and aspects of its historical and current role in the world and also learning about other countries and their historical and current roles in the world. Vital intercultural skills include intrapersonal (or within individual) abilities, such as analysis, emotion regulation, and reflection and interpersonal (or between individual) abilities such as relating and engagement, observation, and listening. Regarding attitudes, the Journeys course focuses on self-knowledge, personal characteristics, and self-efficacy (intrapersonal) and respect, openness, and empathy (interpersonal). These intra- and interpersonal knowledge sets, skills, and attitudes are interpreted through the lens of liberal arts learning, in which intellectual, social, and personal development inform how students situate themselves within a wider context. As active participants in the Journeys course, students work towards learning outcomes including the following intercultural competence-related ones:

• Recognize diversity and inequalities within and among societies in a global context
• Recognize varied perceptions and viewpoints of self and other cultures
• Identify the complexities of intercultural communication by articulating basic cultural norms and behaviours in travel destination.

The Journeys course serves as one of the foundational elements of the multipronged Summit curriculum. The concepts presented in class, embodied in the global study tour, and assessed through multimethod assignments, help first-year students develop basic knowledge, skills, and awareness that will be enhanced in subsequent college years.

All students participate in a multimethod longitudinal assessment called the Global Perspectives Study that takes an intersectional approach to evaluating the process of intercultural competence development and captures how different aspects of students' identity influence this trajectory. In order to meet our goals, our questions about outcomes could not be as simple as "Did change occur during off-campus study, and if so, of what type?" The real usefulness of the data comes in looking at what factors correlate with the degree of change, such as country, program model, duration, student demographic characteristics, and student goals. These comparative or differential impact analyses yield the information that is actually helpful to us as educators, program developers, and advisers. We are also interested in changes in impact over the longer term (as Glass recommends, this volume): Does immediate change disappear over time, stay stable, or even increase as seeds planted continue to blossom?

The Global Perspectives Study starts at baseline (before students arrive at the institution), includes a follow-up survey each year, and terminates with two long-term follow-up surveys given one and five years post-graduation. The instrument collects data on students' (1) basic demographics (e.g., age, gender, race/ethnicity, socio-economic status); (2) home community (e.g., region, racial/ethnic diversity of community and schools); (3) family variables (e.g., languages spoken at home, beliefs about travel and diversity, immigration status, education level); and (4) individual characteristics (self-efficacy, ethnic identity, national identity, cognitive empathy, personality traits, and anxiety/depression). As with all self-report measures, the data can be influenced by various factors including participants' desire to present themselves in a positive light, idiosyncratic response tendencies, or confusion about survey questions.

In addition, the instrument inventories students' holistic development of intercultural maturity on the dimensions of cognitive, intrapersonal, and interpersonal domains using the Global Perspectives Inventory (Braskamp et al. 2008). It also collects data on a range of students' globally related academic, social, and occupational experiences.

The Global Perspectives Study also assesses students' self-reported goals for their international travel experiences (e.g., to have a good time, greater understanding of different cultures). We classify student goals into two broad classes – personal development and intercultural development. Personal development goals include, for example, questions about the desire for greater autonomy, the opportunity to travel, and growth in interpersonal skills. Intercultural goals include such things as the desire to understand different cultures, to learn a different language, and to develop a new perspective on American culture. Surveying goals prior to departure serves pedagogical interests: it has been demonstrated in broader research on learning that when students are primed by reflecting on their own goals, they are more mindful about their encounters, choices, and learning (Paige et al. 2006; Kitsantas 2004). We can also link pre-departure goals to student outcomes, allowing us to recognize that student goals might differ from institutional goals, and student intention has an independent impact on outcomes. In other words, as one might expect, students who are hoping to learn more about world affairs and our place in the world are more likely to show growth on the Global Perspectives Inventory and National Identity Measure than those whose motivations are more about independence and adventure, or even shopping – which students do honestly admit sometimes!

More broadly, the Global Perspectives Study assesses long- and short-term student outcomes (e.g., satisfaction, diversity of friendships, job placement) and institutional outcomes (e.g., alumni engagement, retention, admissions yield). As part of the Global Perspectives Study, we incorporate an in vivo travel observation and self-monitoring protocol to better understand the subjective, emotional experience students have before, during, and after travel. Two weeks before the global study tour, students track their mood using a mobile application throughout the day and report their assumptions about how they believe they would feel while travelling and how engaged they thought they would be with various aspects of the travel experience. During their travels and for two weeks thereafter, students track their mood and report their engagement on a daily basis. In addition, trip co-leaders report their observations of students' mood and engagement levels while travelling. Finally, an assignment for all students in the Journeys course is evaluated by faculty course leaders using a tailored rubric that assesses the key learning outcomes. This rubric translates the wider student learning objectives into a concrete, measureable scale that can be used in several ways. For example, the same rubric can be repeated several times to measure change over time or be compared across students and sections to explore similarities and differences. In addition, students could rate themselves and their peers and/or program directors could use the tool to provide feedback to students. As you can imagine, there are various ways to develop rubrics and utilize the data derived from them. (See Table 8.1 in the chapter appendix for a sample rubric.)

To better understand the impact of these experiences, students and faculty participate in a multitiered assessment process: an assessment of student learning outcomes, evaluation of the course and global study tour, and a broad assessment of outcomes related to global competence development in the higher education environment. Faculty identify the common core learning objectives for the Journeys course and play a key role in assessing the success of students acquiring the learning outcomes through rubric-reviewed assignments and pedagogical assessments. Evaluative data on student satisfaction and concerns help the institution refine the logistical aspects of travel for future students. Finally, wider longitudinal assessment data taken pre-travel (before students enter the institution), and then again post-travel, track several outcomes of interest (e.g., global competence, self-efficacy, academic performance) while considering students' intersectional identities, prior experiences, and individual traits and experiences.

Assessment Challenges

Intercultural competence attained through one specific program, such as Journeys, is difficult to tease out from within a larger multidimensional initiative and requires some creativity. It is not really feasible for students to engage with in-depth assessment tools both immediately before and after a short-term off-campus study program, without suffering from test-retest effects or survey fatigue.

In addition, the gold standard in psychological research on development is a study with a stable control group. With a "condition" such as study abroad, if one wanted to make truly causal claims that changes measured between the pre- and the post-travel measures are caused by study abroad rather than just by maturation, one would also need to measure a "control" group that does not study abroad. This has proved difficult in the field of off-campus study programs, because we clearly cannot randomly assign students to participate or not. We therefore need to be careful, in principle, to avoid language of causation (e.g., claiming that study abroad causes x or y) and instead talk about correlates of change (those who studied abroad are also more likely to x or y.)

One of the major limitations that we have repeatedly come up against (and which may be related to the capacity concerns raised by Owens and Sotoudehnia in their chapter in this volume) is the amount of time that some assessment procedures take, and especially qualitative methods. While we are often interested in gathering student responses in a more unfiltered way, either from application essays or classroom assignments, we have found the amount of time needed to analyse this data stream unattainable in light of the many more immediate tasks that are required of our office. As a result, we have relied on mostly quantitative assessment procedures that can be collected electronically, but we do continue to try to find ways to work in at least some of the more labour-intensive qualitative responses.

Finally, sustaining overall response rates can be a challenge, even though students are usually intrinsically interested in reflecting on these experiences. In order to encourage responses, we have used several types of incentives (e.g., gift card drawings) with some success and have also brought onto the research team qualified faculty and students who may have more of a relationship with students. We have had psychology students getting credit for independent studies or internships join us, as well as others with a professional interest in the research. Of course, we also draw on our own institutional research office as appropriate.

Developing intercultural competence among college students is a complex, ongoing, and multifaceted process. Accordingly, assessment is particularly challenging. To address this challenge, assessment techniques must be able to incorporate various elements, including the broad range of experiences that individuals bring with them to the higher education environment, social influences, and specific individual and demographic considerations.

Concluding Thoughts

What we are trying to capture is nothing less than the personal and intellectual transformation our students so often experience in the context of off-campus study programs. This is completely different from student satisfaction, which may be worth evaluating in its own right. Carefully envisioning the potential transformation and choosing methods well can yield results that help not only to describe it, but also to sustain it, enhance it, and share it with others. There are many challenges to high-quality critical assessment, but given the enormous resources we invest in creating learning contexts "out there," basing these on hunches, good intentions, or vague "aha" moments is clearly not adequate. We need to assess to what extent students achieve both their own transformational goals (whether academic and/or socio-emotional) and ours as an institution for ourselves or our host partners, so that we can better assist future students in doing so.

Appendix

Table 8.1 Journeys Rubric with Selected Learning Objectives

Learning Objective	Advanced (4) Many strengths present, few weaknesses	Competent (3) On balance the strengths outweigh the weaknesses	Developing (2) Strengths and need for improvement are about equal	Beginning (1) Need for improvement outweighs strengths
Recognize varied perceptions and viewpoints of self and other cultures.	Varied perceptions/ viewpoints are presented clearly and described comprehensively, providing all relevant information necessary for nuanced understanding.	Varied perceptions/ viewpoints considered and stated so that understanding is not seriously impeded by omissions.	Perceptions/ viewpoints considered, but without recognition of variety or multiplicity; leaves some terms undefined, ambiguities unexplored, boundaries undetermined, and/or backgrounds unknown.	Perceptions/ viewpoints are unstated or stated without clarification or description.
Draw connections between travel destination/ experiences and course concepts.	Incorporates specific course concepts (as expressed in course objectives, course readings, and/or class discussions) in ways that are clear, with synthesis of content and experience.	Incorporates specific course concepts (as expressed in course objectives, course readings, and/or class discussions) in ways that integrate concepts with experience, but misses opportunities for depth and/ or coherence.	Articulates specific course concepts (as expressed in course objectives, course readings, and/or class discussions) in ways that do not demonstrate understanding or connect with experience.	Does not articulate specific course concepts (as expressed in course objectives, course readings, and/or class discussions) or connect to experience.

Learning Objective	Advanced (4) Many strengths present, few weaknesses	Competent (3) On balance the strengths outweigh the weaknesses	Developing (2) Strengths and need for improvement are about equal	Beginning (1) Need for improvement outweighs strengths
Recognize diversity and inequalities within and among societies in a global context.	Describes diversity and inequalities with sufficient interpretation/ evaluation to present a comprehensive recognition.	Describes diversity and inequalities with sufficient interpretation/ evaluation to develop a coherent recognition.	Describes diversity and/ or inequalities with some interpretation/ evaluation, but not enough to develop a coherent recognition.	Does not describe diversity and/ or inequalities with sufficient interpretation/ evaluation.

NOTES

1 For ten more case studies of practice in assessing study abroad, see Savicki and Brewer's excellent recent edited volume (Savicki and Brewer 2015).
2 This core includes an initial leadership development immersion, a four-credit Leadership Prologue course and one-credit Global Gateways course in the fall of their first year, and then a one-credit Leadership Dialogues course and four-credit Global Journey course (which is the focus of this case study) in the spring. All students also take sixteen credits of a language, three more distributional requirements which are infused with Global Learning and Leadership Development, and choose to specialize in one of these themes in addition to their major. These specializations include more courses and co-curricular activities, such as another study abroad experience for the Global Learning one and a week-long Team Global Challenge competition.

REFERENCES

Anderson, Phil, Leigh Lawton, Richard J. Rexeisen, and Ann C. Hubbard. 2006. "Short-term Study Abroad and Intercultural Sensitivity: A Pilot Study." *International Journal of Intercultural Relations* 30 (4): 457–69. https://doi.org/10.1016/j.ijintrel.2005.10.004.

Braskamp, Larry A., David C. Braskamp, Kelly Carter Merrill, and Mark Engberg. 2008. "Global Perspective Inventory (GPI): Its Purpose, Construction, Potential Uses, and Psychometric Characteristics." ResearchGate. Accessed 15 Dec. 2016. https://www.researchgate.net/profile/Larry_Braskamp/publication/239931705_Global_Perspectives_Inventory_GPI_Its_Purpose_Construction_Potential_Uses_and_Psychometric_Characteristics/links/542193750cf203f155c6e305.pdf?origin=publication_list.

Chieffo, Lisa, and Lesa Griffiths. 2004. "Large-Scale Assessment of Student Attitudes after a Short-Term Study Abroad Program." *Frontiers: The Interdisciplinary Journal of Study Abroad* 10: 165–77.

Cohen, Jonathan. 2006. "Social, Emotional, Ethical, and Academic Education: Creating a Climate for Learning, Participation in Democracy, and Well-being." *Harvard Educational Review* 76 (2): 201–37. https://doi.org/10.17763/haer.76.2.j44854x1524644vn.

Deardorff, Darla. 2015. *Demystifying Outcomes Assessment for International Educators: A Practical Approach.* Sterling, VA: Stylus.

Institute of International Education. 2016. "Open Doors Report on International Educational Exchange." Accessed 17 Sept. 2017. https://www.iie.org/Research-and-Insights/Open-Doors/Data/US-Study-Abroad/Duration-of-Study-Abroad.

Kehl, Kevin, and Jason Morris. 2008. "Differences in Global-Mindedness between Short-Term and Semester-Long Study Abroad Participants at Selected Private Universities." *Frontiers: The Interdisciplinary Journal of Study Abroad* 15: 67–79.

Kitsantas, Anastasia. 2004. "Studying Abroad: The Role of College Students' Goals on the Development of Cross-cultural Skills and Global Understanding." *College Student Journal* 38 (3): 441–52.

Moeller, Aleidine Kramer, Janine M. Theiler, and Chaorong Wu. 2012. "Goal Setting and Student Achievement: A Longitudinal Study." Faculty Publications: Department of Teaching, Learning and Teacher Education. Paper 159. Accessed 15 Dec. 2016. https://digitalcommons.unl.edu/teachlearnfacpub/159. https://doi.org/10.1111/j.1540-4781.2011.01231.x.

Paige, R. Michael, Andrew D. Cohen, Barbara Kappler, Julie C. Chi, and James P. Lassegard. 2006. *Maximizing Study Abroad.* Minneapolis: University of Minnesota Press.

Savicki, Victory, and Elizabeth Brewer. 2015. *Assessing Study Abroad: Theory, Tools, and Practice.* Sterling, VA: Stylus.

Turkay, Selen. 2014. "Setting Goals: Who, Why, How?" Manuscript posted on website of Harvard Vice Provost for Advances in Learning. Accessed

15 Dec. 2016. https://vpal.harvard.edu/publications/setting-goals-who-why-how.

Vande Berg, Michael, Michael R. Paige, and Kris Hemming Lou. 2012. *Student Learning Abroad: What Our Students Are Learning, What They're Not, and What We Can Do About It.* Sterling, VA: Stylus.

9:14 AM: SATURDAY, 14 MAY 2016

I have stumbled on the bones of broken corpses
Made love with the ghosts of parchment paper

Wondered and wandered
Walked in courtyards of a broken past

Wanted to repair myself

But found it wasn't I that was in despair

Seen unmarked graves
And empty plots

Exiled from the garden

Forced to settle in the ash
To hear the Kaddish in the streets of East Berlin

Jake Noah Sherman
Participant, University of Victoria's
I-witness Holocaust Field School

"THE WORLD MOVES THROUGH US" ...

This is the story of the encounter that inspired me to begin asking questions about my heritage and get on a night bus to Lithuania, a place I never thought I'd go.

I saw her at dawn. Her brown curly hair blew in the wind on a cold Polish morning. She'd been looking at me the whole day previous at Jagellonian University during our joint class session.

Walking up to me with confidence, and readjusting her glasses, she looked me right in the eye: "I think I know you."

"I don't think so," I answered curtly, "but nice to meet you."

She shook my hand, gave me a look equal parts bewildered and embarrassed, and began to walk away. I grabbed her shoulder as she turned: "My name's Jake Noah."

I got another puzzled stare and a coy smile.

She laughed, "Maybe we have met."

I responded bluntly with a hint of sarcasm this time, "I don't think so, but who knows, maybe in another life."

She nodded her head, "Did you know your middle name in Hebrew and my last name in Polish mean the same thing?"

"No. And what's that?"

"Comfort, or comforter," she said. "My name's Shifi. Shifi Wygoda."

We boarded the bus and took our seats in silence. I looked at her mystified: "How'd you know that?"

"I studied linguistics in Israel."

"And what are you doing in Poland?"

"Investigating my roots, learning the language. I was born in Montreal, but grew up in Israel and have come to find that Poland is really where I come from."

We spent the rest of the day together, most of it calm and quiet, simply enjoying the comfort of one another as we wandered in silence towards mass graves and paid our respects.

On our way home we stopped for gas. She and I sat alone in the grass.

She turned to me: "You know, in Hebrew, the word for person, 'adam,' is necessarily singular. The word for life, 'chayim,' is necessarily plural." I turned towards her slightly. She leaned in. "The world moves through us," she whispered.

We boarded the bus and continued moving East in silence. I paused. Tried to listen to my breath, but couldn't hear it over the beating of my heart.

Two weeks later I found myself on another bus, watching the sunrise over a crescent moon in the land of my ancestors: Lithuania. And all of it, all I found out in the coming days, wandering into synagogues and through tiny former ghetto streets, was because of Shifi. The world continues to move through me, because of her.

Jake Noah Sherman
Participant, University of Victoria's
I-witness Holocaust Field School

9

Assessing Learning "Out There": Four Key Challenges and Opportunities

CAMERON OWENS AND MARAL SOTOUDEHNIA

An Invitation to Assess Field Learning

For years, geographers and other scholars have urged students out of the classroom and into the so-called real world to study (Fuller et al. 2006; Sauer 1956). As this volume makes clear, field schools or study abroad programs[1] are celebrated for providing enjoyable, experiential, relational, and place-embedded learning. However, while there is a widespread sense that these are valuable opportunities, and students often return gushing about their experiences, there is pressure to examine the impacts of these programs more intentionally (West 2015, 1).

Administrators, for instance, demand confirmation of the value of travel study to justify continued investment in programs that are often expensive and have lower enrolment than classroom-based courses. Institutions offering field schools "face a demand for aggregate data that demonstrates both program results and student growth" (West 2015, 1). Where the impetus for evaluation is not driven by administrative concerns of accountability, instructors may be moved to evaluate programs as part of the good practice of continuous improvement, perhaps with an uneasy sense that the impacts of such programs are not always positive. Indeed, field study has been the subject of critical questioning around privilege and positionality, the colonial and masculinist assumptions undergirding these excursions, their environmental and social impacts, and their status as "entertainment paradigms" (Castleden, Kiley, Morgan, and Sylvestre, this volume; Gray 2015; McMorran 2015; Nairn 2005; Shurmer-Smith and Shurmer-Smith 2002; Tiessen and Hulsh 2014). One student at Concordia College in Minnesota described study abroad as a "privilege reserved for the global elite" that perpetuates imperialism and classism, involves serious non-negotiable environmental costs, encourages the dismissal of local issues (i.e., one does not need to leave the United States to study poverty, racism, and other problems), and facilitates "mining for knowledge and walking away while the people I'm learning about continue to suffer" (Gray 2015, paras. 5–14). While in their chapter in this volume, Peifer and Meyer-Lee point out that "at a historic moment of increasing nationalism and isolationism" international field learning may serve an urgent need, the critical concerns raised by this student invite attention.

In a similarly critical vein, Nairn (2005) disputes the popular binary, implicit in field study, between a "real" world out there affording direct experiences and a world of campus libraries and classrooms and their supposedly disconnected theory (Nairn 2005, 293). Reflecting on a field program in New Zealand, she challenges claims that "first-person research" – in this case, talking with migrants in their own setting rather than just reading about them in the classroom – provides students with unmediated insight into the experience of migration and race, or protects them from the inductive influences of their preconceptions. In her case, students' preconceived racial differences and fears of the "other" seem to be merely confirmed by field observation. For Nairn, ideology insidiously shapes perception even for students engaged in direct field observation, reinforcing misconceptions and unintentionally reproducing the social setting the program intends to discredit. Accepting this argument, if students are not made mindful of their own positionality or preconceptions, and instructors do not actively attempt to disrupt such perspectives, field programs risk reaffirming problematic world views. While Hope (2009) objects to the inevitability implied in Nairn's account, calling for a more variegated understanding of student experiences, such a cautionary note represents compelling motivation to intentionally and comprehensively evaluate the impacts of field study.

Not surprisingly, the task of assessing such programs is far from straightforward. Critical scholarship has revealed that assessment, in a variety of forms – for instance, environmental impact assessment, economic forecasting, or risk analysis – is confounded by complexity, uncertainty, and the challenge of measuring what is often unmeasurable (Stone 2013; Pilkey and Pilkey-Jarvis 2009; Gibson et. al. 2005). The prospect of assessing learning is every bit as enigmatic. It is unlikely that simple, direct causal links can be detected between teaching and learning, calling into question the possibility of unequivocally identifying whether, when, how much, for how long, and where learning takes place (Ellsworth 2005;

Hussey and Smith 2003; Hern 2016). Moreover, assessment is not an entirely neutral or technical exercise but is inescapably political, set in a context of competing assumptions, values, and power asymmetries (Stone 2013). There is little consensus in higher education as to the ultimate aims of learning in the first place, let alone the means for evaluating it. The goal of one field program, for example, may be to produce savvy individuals well positioned to compete in a fierce, global economy, while another may seek to cultivate citizens critical of the normalization of the very discourse of competitiveness. One program may seek conceptual clarity and discipline-specific skills development, while another may seek to challenge dominant discourses and disrupt disciplinary boundaries, and so on.

In this chapter, we attempt to make sense of formidable challenges surrounding the assessment of field programs. Our[2] insights arise out of reviewing the nascent literature on field school evaluation, engaging with other field school instructors,[3] and reflecting on experiences running and assessing our own program over the past six years. The annual, month-long University of Victoria (UVic) Sustainability Field School affords undergraduate students the opportunity to travel and connect with urban planners, activists, and other community leaders grappling with the daunting socio-ecological challenges facing cities in the (North American) Pacific Northwest or (European) Atlantic Northwest. Evaluating outcomes represents a significant challenge that we have approached through longitudinal studies employing a range of qualitative methods and a spirit of experimentation[4] (see also Owens, Sotoudehnia, and Erickson-McGee 2015).

Providing generalized guidance for evaluating field schools is a daunting undertaking, given the wide range of field programs, locations, aims, and objectives (see Peifer and Meyer-Lee, this volume). We are quick to acknowledge the limitations of our intervention. We do not pretend that these observations constitute a comprehensive overview of all the forms of assessment or types of field learning situations; see Fuller et al. (2006) for an overview of the assessment literature. Rather, we have flagged what, in our experience, are crucial considerations in the spirit of advancing the conversation and supporting others who are struggling to unpack what happens when students venture to learn "out there." We are sympathetic to the "assessment society" critique (Broadfoot and Black 2004, 19), that our culture is unhealthily obsessed with numbers, targets, and accountability often linked to neoliberal priorities of efficiency, competitiveness, and profit. Our own call to assessment here is not primarily motivated by institutional concerns with accountability (although those pressures exist), as with our own curiosity (and healthy apprehensions) about the impact of our teaching.

In what follows, we consider the following four specific challenges that may confound the evaluation of field programs: (1) building the *capacity* to marshal the resources to undertake assessment in the first place; (2) developing an adequate *scope* (i.e., parameters of study), an issue we consider at some length; (3) using appropriate *methods* to obtain trustworthy data; and (4) *identifying learning* given the politically complex nature of the subject matter in the often unscripted context of field programs.

Building Capacity

A prefatory challenge in evaluating field study surrounds capacity. Most universities institute some kind of standardized evaluation process for all courses, including field offerings. However, these are usually fairly blunt tools for assessing learning and are often regarded as being more about accountability or popularity (Emery, Kramer, and Tian 2003). Institutions or individual instructors may seek more nuanced means for evaluating programs, but these require time, will, expertise, and resources. Figuring out how to marshal resources (time, money, know-how) in today's climate of austerity is often a significant challenge.[5]

Nevertheless, opportunities do exist to disencumber evaluation. With the current buzz of experiential or international education, resources are sometimes available through campus teaching and learning centres to support the development and evaluation of such programs. For instance, our own Learning and Teaching Centre assisted in the formation of a multidisciplinary support group for field school instructors. Through hosting workshops, creating an online guide with practical advice for travel study planning, and initiating this book project, this group helped address the isolation and burden of developing and assessing field programs. It has been a useful forum for sharing good practices, and we encourage other institutions to consider such a means of capacity building.

Designing assessment right into the field program represents an effective opportunity to unburden the process of evaluation. Instructors, potentially in coordination with students, can plan individual and group reflective exercises that, in turn, provide rich qualitative data for analysis (Elwood 2004; Glass 2014). In our case, qualitative data are obtained through required course elements, including a field journal (capturing

daily observations, reflections, and connections), frequent group sharing circles, and a reflective assignment. The latter exercise affords particularly useful insights into student learning. As an initial course activity,
students are asked to articulate their preconceptions of the course focus,
"sustainable community development," to formulate key questions that
will guide their inquiry in the field, and to identify their specific learning expectations. We surprise the students on the final day of the course
when we produce these original musings and have them revisit and
reflect on them. Such an exercise encourages explicit engagement with
and recognition of field learning, serves as part of student's course grade,
but also provides useful data for our overall assessment of the program.

Evaluation can be made less burdensome on instructors by involving students, which in turn supports a more student-centred learning
model. For instance, as part of a directed studies course following our
2013 field school, one of our alumna undertook follow-up interviews
with her course-mates and documented, transcribed, coded, and analysed their learning experiences. This research contributed to developing
an assessment rubric, a documentary video on field school learning,[6] and
a publication (Owens, Sotoudehnia, and Erickson-McGee 2015). This
exercise in triangulation (or cross-verification) provided unique student-
centred reflections on the value of course elements that may otherwise
have been overlooked.

Scoping the Assessment

Beyond capacity, assessors face the challenge of deciding what to include
in an assessment. In environmental impact assessment this step is called
"scoping," whereby one determines what to consider in a review: questions to ask, variables to track, criteria to use, and tools to employ. The
range of elements that could be considered in the evaluation of a field
program is potentially endless. However, the scope of a review is constrained by capacity and guided by the assumptions, values, imagination,
and awareness of the assessor. Identifying parameters in any context,
field study or otherwise, is challenging. Approaching learning in terms
of complex adaptive systems theory, Burns and Knox (2011) identify an
array of integral factors, including elements they classify as institutional
(e.g., time pressures, required materials, course syllabus), pedagogical
(e.g., previous learning, student needs, teacher-student relationships),
personal (e.g., personal lives and relationships), and physical (e.g.,
level of comfort, physical size and layout of class, student movement).

Imagining learning in such a way hints at the challenge of comprehensive assessment.

While we reject setting up a rigid binary between conventional and field learning (see Borrows, this volume; Musisi, this volume), there are a number of factors that are particularly, if not uniquely, relevant to learning "out there." As other chapters in this book have introduced, field programs can, among other things, foster awareness of the situated nature of knowledge, encourage thinking relationally, foreground the complex dynamics of places, encourage adaptation to unfamiliar environments, and cultivate intercultural literacy. Attuning evaluation to consider such elements extends beyond most conventional assessments. In their chapter in this book, Peifer and Meyer-Lee identify a number of useful questions to ask and tools to employ in evaluating learning "out there." Their own scoping involves attending to concerns with academics, students' socio-emotional life, the host community, and the home institution. Complementing their work, we consider a range of elements that a comprehensive evaluation of field learning could take into account, including *situational elements* such as itinerary and group dynamics, *subjects* beyond just the participating students, the *broader disciplinary program* within which a particular field school is embedded, and the *environmental and social impacts* of the field course.

Our research has attended to a number of situational elements, many unique to field excursions, that can influence the overall quality of the learning experience.[7] For example, the itinerary of a travel study program can have a profound impact on learning in terms of the relevance of the itinerary locations and the strength of the narrative thread weaving the trip together; time allotted to each itinerary location; balance between structured and unstructured time; the location and quality of accommodations; and the extent to which the instructor makes explicit and compelling the justifications for decisions about each of these.

Given the immersive quality of travel study, cohort or group dynamics appear relevant and should be on the field school developer's radar screen. Factors that may influence learning include the relationships among instructor, assistants, and students; personalities constituting the cohort; energy levels and morale; quality of the discursive environment (e.g., the level of comfort students have in expressing themselves); skill level of the facilitator; and relationships with local community hosts. To the extent that learning is a social process, the relevance of group dynamics is increasingly attended to in assessment. As our field school has progressed over the past six years, we have become more attuned to the collective learning taking place. In follow-up interviews, students frequently reflect that they learn as much from each other, through sharing circles and informal engagement, as they do directly from course activities or the instructor. We have encouraged students in our program, which studies sustainable community development, to imagine ourselves as a little community. It has not been lost on some students that our own group represents a microcosm of the broader societal processes we are studying and that we can learn through our own issues, relationships, and collective (mis)understandings. Extracurricular activities, such as communal cooking, take on an important role not only with respect to the development of life skills, but also in revealing social relationships of value to understanding community. This recognition, arising out of earlier field school assessments, has informed how we structure our current excursions. We now spend much more time explicitly considering group dynamics and exploring with students how such dynamics influence their learning. Increasingly, we include students when planning the field school program and in the learning assessment itself. Students interview each other and even help to develop and refine flexible learning outcomes.

While student learning is usually the focus of assessment, we need also to acknowledge that the impacts of field programs extend beyond the participating students, inviting us to widen the scope of evaluation. For example, a comprehensive assessment could track:

- Impacts in terms of professional development in research and teaching for both faculty and graduate student teaching assistants
- Impacts on other students not on the field school but who take courses from instructors or with students who participated
- The physical and emotional toll on instructors given the 24/7 nature of the course itself, including extensive and often unrecognized logistics planning, and the level of responsibility for caring both for the students and the communities and organizations with which the students engage
- The field program's contributions to the learning goals and reputation of the university
- The long underappreciated impacts on communities who host programs with or without consent (as discussed in detail here in the chapters by Vibert and Sadeghi-Yekta and by Castleden, Kiley, Morgan, and Sylvestre).

In broadly considering the participants of field schools, we may also fine-tune assessments to attend to gender, race, sexuality, ability, and other markers of identity, which can reveal some of the variegated impacts of programs (Vibert and Sadeghi-Yekta). As one illustrative example, Hughes (2016) studied how field trips to rural Britain were imagined differently by white British students and British students of colour. Such "domestic" trips were assumed to be safer and more comfortable, requiring less preparation than study abroad. However, such excursions provoked high levels of anxiety among the students of colour, responses overlooked by field school organizers in the context of normative whiteness. As part of our program, we invite our campus Anti-Violence Project to host a pre-session event, where topics such as privilege, consent, and forms of discrimination and violence are raised with our students and the instructors. We would highly recommend this inclusion in other programs.

Another consideration relates to a recognizable trend whereby institutions have sought to expand the scope of assessment beyond individual field courses to examine a range of courses on offer and/or to attend to how a field course integrates with and supports broader departmental or even university-wide goals (West 2015). This more extensive scoping involves considerable coordination, a capacity challenge to which few institutions have risen (although see Brewer and Cunningham (2009) on integrating study abroad into the undergraduate curriculum). It is an important frontier to consider, inspiring attention to how field schools extend rather than replace or ignore the development of theory in the classroom (Fuller et al. 2006; Nairn 2005). As Shurmer-Smith and Shurmer-Smith (2002, xx) argue, "A field class should not just be a thing in itself, it should act as the fulcrum of other learning."

Whether through individual courses or as suites of programs, comprehensive assessment should encompass social and environmental impacts. Student researchers at Concordia College in Minnesota, for example, revealed that study abroad programs were responsible for 15% of their institution's greenhouse gas emissions (as referenced by Gray 2015, para. 6). To the extent that the ostensible goal of many field schools is to inspire more socially and environmentally aware citizenship, an honest, explicit, and thorough accounting of such potential effects is imperative. Otherwise, the unintended pedagogy is one of mixed messages and hypocrisy. Indeed, in this time of global crisis, Gray (para. 7) suggests we shift from asking why we want to study abroad to asking whether we need to. If we do continue to offer programs "out there," instructors need to be sensitive to potentially problematic colonial, racial, legal, or other dynamics with respect to the places they visit. In our province of British Columbia, where sovereignty over the land is contested between the colonial state and Indigenous First Nations, no excursion "out there" is free of ethical, political, and legal dynamics. Indeed, these are not places "out there" for the host communities. These dynamics may play out differently in different locations, but having the goodwill and consent of the communities we visit must be at the forefront of the assessment (and delivery) of field programs.

Given the substantive focus of our field school on sustainable community development, we are particularly concerned with residual social and environmental impacts. On our itinerary, we meet with Indigenous leaders, planners, community organizers, academics, and others grappling with critical urban socio-ecological challenges. En route, we have open discussions about our own privilege and consumption, travel, and waste habits. Along with gifts and generous remuneration, we acknowledge the community partners who share their time and knowledge with us through efforts ranging from weeding a community garden to producing and sharing a documentary video to showcase a local project. We try to make the impact of our travel explicit throughout the course, and we challenge ourselves to think of creative ways to offset those impacts. To this end, and to build on the momentum of the field school, students undertake mandatory "legacy projects." These projects involve sharing what we have learned on the field school and acting in our home communities and take a variety of forms, such as making presentations to local councils or schools, initiating grassroots sustainability projects, or undertaking research or outreach work valuable to a local community partner. In assessing our program, we follow up with each presenter asking about their experience participating to ensure mutual benefit. Still, the effort to precisely measure impacts, mitigate negative impacts, and ensure the acceptability of trade-offs remains a significant concern.

One final example illustrates the value of broadening the scope of review. In the assessment of our 2015 trip, we discovered that students were concerned about their inability to navigate in foreign environments. The instructors had taken care of all the transit arrangements, and students then faced a steep learning curve travelling onward after the field school had ended. This practical consideration and valuable learning opportunity had been missed. In response, in the 2016 version of the field school, we tasked groups of students with navigating transit systems and train schedules and getting us around in a timely fashion.

This small adjustment had meaningful impacts that many students noted in their follow-up assessment, such as the development of leadership skills and the demystification of non-automobile transportation, which in some instances seems to have translated into behavioural change at home. This shift illustrates the value of self-guided learning and the benefit of broadening the scope of evaluation.

In this section we have drawn attention to a wide range of elements that might be ideally considered in a comprehensive evaluation of a field school. Of course, this is not an exhaustive list, nor are we suggesting that instructors will always have the capacity to assess all of these dimensions. Our hope is that this consideration encourages assessors to think broadly about (and intervene in) the potential impacts of programs. To this extent, we can see assessment as less about simple measurement and more about expanding our radius of concern or attention. It is also worth noting that students, teaching assistants, community members, and others can be sources of valuable insight.

Producing Reliable Data

Carefully identifying what to include is crucial to a valuable assessment, yet an additional challenge involves actually undertaking the evaluation – obtaining or producing[8] high-quality data. The positivist faith in the ability to precisely quantify complex social phenomena, such as learning, has long been discredited in the social sciences (Stone 2013). If assessing learning were simply about assigning numbers to indicate content mastery, conceptual clarity, and manual skills development, measuring competence might be more straightforward. However, learning, especially in the contentious social settings of the field, is a much more complex phenomenon. A variety of survey tools have been employed to try to capture various dimensions of field learning, especially with respect to intercultural dynamics[9] (see the chapter by Peifer and Meyer-Lee). Such aggregate-level assessments are indeed useful, but in-depth qualitative methods provide more nuance for assessing such elements as the extent to which a student understands the situated nature of learning or can think relationally.

In our own evaluations, we have used reflective assignments, semi-structured interviews, focus groups, and even surveying social media to get a sense of the impacts of our program. Of course, volumes have been written about the challenges of such qualitative analysis (Baxter and Eyles 1997; Kirby, Greaves, and Reid 2006; Silverman 2015). It is impossible to adequately consider all such concerns in this short chapter. In what follows, we flag specific challenges we encountered in analysing student interpretations of their learning.

Asking students to explicitly reflect on their learning can be a useful and revealing starting point, but obtaining trustworthy data is a demanding task (Baxter and Eyles 1997). It can be difficult for students and teachers to identify what counts as a moving learning experience. In interviews with students, the problem of the double hermeneutic recognizes that the researcher, investigating through her own social lens, must interpret the students' interpretations of their learning (Giddens 1987). The following excerpt from Nairn's (2005) critical evaluation of a field course in Kenya designed to challenge preconceptions, ignorance, and stereotypes, illustrates concern with the trustworthiness of interpretation.

> The students ... celebrated direct experience, for example, "Experiencing Kenyan culture first hand was amazing" ... but it is very difficult to glean any sense from the data presented that stereotypes might have changed as a result of the field course. For example, one student claims that this "unbelievable academic, and personal experience ... will no doubt change my way of thinking for the rest of my life" ... but, as readers, we do not know how this student's thinking has been changed. In another case, a student comments that "seeing the poverty first hand instead of on TV is really upsetting"... but how this affects this student's preconceptions is left to the reader to ponder. (Nairn 2005, 296)

The student reflections suggest a transformative experience, but the claims are largely unsubstantiated. Nairn (2005) cautions against the assumption that "students' accounts provide transparent meaning concerning transparent experiences," which entraps us in "a referential notion of evidence" (297). Taking these concerns seriously invites us to pierce through the excitement of the program to seek more substantive evidence of learning and to investigate behavioural change.

One immediate challenge surrounds the incredibly complex relationship between emotions and learning (Pekrun and Linnenbrink-Garcia 2014). The researcher needs to be circumspect about students gushing about the great (or horrible) time they had on a field school, working through (without dismissing) the emotional experience to find evidence of learning. "Positive" experiences do not automatically equate to deep learning (although they might). Conversely, "negative" experiences – moments of personal or group discomfort or frustration – can be powerful catalysts of growth (think of Mezirow's (1978) "disorienting dilemma," referenced in Glass, this volume) but are often misrecognized, at least immediately (see Vibert and Sadeghi-Yekta, this volume). Evaluations of a program could be correlated with spurious factors such as group dynamics, the weather, cleanliness of accommodations, or even the quality (or quantity) of local beer. Complicating matters, it is impossible to parse out the emotional and get to some pure, rational core of the learning that took place (Pekrun and Linnenbrink-Garcia 2014). The extent to which students are excited about their learning can be revealing of the effectiveness of the program, but a more sustained analysis is needed to confirm the depth of learning (see Peifer and Meyer-Lee's consideration of socio-emotional transformation in their chapter in this volume).

Another related challenge surrounds determining the optimal timing of evaluation (see McKinney, this volume). Logistically, it is easiest to undertake evaluation immediately at the end of a program. At this point learning may be at its freshest, but it may be too early for many students to fully articulate how they have been affected. Determining the depth and longevity of impact is further complicated by the diverse aptitudes of students. To the extent that learning is a lifelong process (Dewey 1916), it becomes challenging for assessors to identify the long-term impact of any particular program on diverse learners (Mitussis and Sheehan 2013). The field school is one experience among many shaping the lives of the students. Longitudinal studies that take snapshots of student learning at different times offer some promise. In 2016, we undertook a follow-up evaluation of the field schools run over the previous four years. More than the length of time that had transpired since their participation, life circumstances seemed to be the most important factor, in terms of student recognition of learning. Students were most insightful and articulate about the specific impacts of field learning when the learning had become directly relevant to their job, volunteer work, or further scholarly pursuits, regardless of the time that had elapsed since their field experience.

One additional consideration surrounds the question of who leads the assessment. Who asks the questions and who else is in the room influence how students will respond. In our case, the principal instructor has undertaken much of the assessment. However, two evaluations, in 2012 and 2014, were undertaken by students. While we are constrained in what we can claim based on non-ideal conditions (e.g., lacking a control group), there seem to be important differences when analysing student-led and instructor-led assessments. While students may lack experience in undertaking semi-structured interviews, they can unearth insights elusive to the instructor, given the power relations inhering in the student-teacher relationship. Some students let their guard down and were more willing to share their frustration or confusion around their learning when talking to a peer. In interviews with the instructor, responses seemed more formal and were more explicitly linked to structural elements of the course, each with its own benefit to understanding the dynamic of the course.

Returning to our concern with the potential lack of evidence in student responses, indications of behavioural change as well as reflexivity (recognizing one's own intimate connection to the subject matter studied) form central concerns in our assessment. Beyond vague statements about the "amazing" experience, we seek evidence of students seeing themselves differently and acting differently in their communities. Useful data were obtained in a 2016 survey where we asked students from four previous field schools to describe the most profound impact of the trip and whether and how they had applied what they learned "in the field" in their professional and personal life. Evidence of the impact of the program showed up in student reflections on how learning in the field had directly led to increased confidence and recognition of how to interact with professionals (e.g., planners, community leaders); involvement in local community projects; political activity; daily habit change; using connections established during the program; and imagining themselves in novel ways (e.g., actually being able to see themselves in certain employment or activist roles).

A tool that has emerged as surprisingly useful in our assessment is a closed Facebook group we have set up for each field school. Social media is used before and during the program for the group to communicate and share images, insights, and valuable links. We encourage the group to continue communicating after the field school, and it is here that we get some of the clearest insights into the impact of the program. Posts reveal student engagement, participation in projects, reminiscences, evolving interests, and employment. The Facebook group itself constitutes an important medium through which our learning community is nurtured and sustained.

Identifying Learning in the Context of Complexity

One more challenge surrounds analysing learning in the context of complexity, both in terms of the political nature of the subject of sustainable community development and the nature of the field setting. Field programs, while clearly curated, remain more or less open to the unintended, unscripted, and spontaneous. This unintentional learning in relatively uncontrolled public places is regarded as one of the most valuable elements of the field experience (Stenglein and Mader 2016). These programs facilitate what Hussey and Smith (2002, 229) see as a key to effective higher education, "the emergence of ideas, skills, and connections, which were unforeseen, even by the teacher." For Guinness (2012, 330), field learning can challenge students to unpack a problem "not from the luxury of reading a literature which has already dissected the issue but from the standpoint of a population 'on the ground' who do not themselves have a consensus on what the issue is." As such, students are invited to be "attentive, patient and consultative in attempting to understand the issues they recognize" and not to jump to "trite conclusions and banal solutions" (330).

The following is an illustrative example of the value of unscripted learning and political complexity. During our 2012 Cascadia field school, we visited the Eco-Flats project, a bicycle-oriented development in northeast Portland, Oregon. Our tour of this celebrated example of sustainable development, with impressive eco-technical features, was unexpectedly interrupted by a community activist who objected that the project represented unwelcome gentrification and part of an ugly process of racist displacement of the Black community in the area (see Hern 2016). This experience disrupted students' understandings of green building and challenged them to recognize the opaque distribution of costs and benefits of different development scenarios. Follow-up assessments, even four years later, revealed that this event had left a deep impression on students and had (de)pressed them to be much more sceptical about "sustainability" and what it masked in its near-hegemonic acceptance (Gunder 2006).

The challenge remains, then, to set up a course structure flexible enough to allow students to benefit from such spontaneous disruptions and open to the uncontainable complexity of social reality, but with some kind of touchstone to guide and assess learning. Rigid adherence to pre-ordained learning outcomes may compromise such opportunities (Hussey and Smith 2002, 2003, 2008). We highlight here our efforts towards such a flexible approach.

In teaching about sustainable community development, we have leaned on a framework introduced to us by one of our field school contacts in 2014, John Robinson at the Centre for Interactive Research in Sustainability at the University of British Columbia (see Table 9.1). This framework describes four sets of attributes and specific learning outcomes to

Table 9.1 A Summary of UBC's Sustainability Education Framework

Student Sustainability Attributes			
Holistic systems thinking: Everything is connected	Sustainability knowledge: Understand the context, know the challenges	Awareness and integration: Connect what I know with what you know	Acting for positive change: Contribute to co-creating a better future
Sustainability depends on, and aspires to, a purposeful, equitable, and harmonious integration of human and natural systems. Holistic, ecological, or synergistic systems thinking (aka "joined-up mindset") requires students to acquire the means and methods to see, articulate, and qualitatively and quantitatively measure how human and natural systems work and interact.	Sustainability depends on comprehensive knowledge within one's area of study, and an understanding of contemporary sustainability issues (particularly those that relate to their own area of study). Sustainability knowledge also requires students to gain proficiency in the underlying ideas and principles of sustainability, and in the evaluation of different sustainability models and paradigms.	Sustainability depends on a broad range of interdisciplinary, cross-disciplinary, and transdisciplinary approaches to allow for the emergence of new domains of knowledge equal to the task of addressing global issues. Awareness and Integration requires students to think and act in new ways to solve complex, integrative problems through collaboration between disciplines.	Sustainability depends on change agents who have the skills, persistence, and resilience to contribute to the emergence of healthy ecosystems, social systems, and economies. Acting for positive change requires students to engage others and implement or contribute to positive change.
Example learning outcomes: 1. Demonstrate a capacity to appreciate all actions have consequences within, between, and among systems.	Example learning outcomes: 1. Compare and contrast three- and four-component sustainability models, and assess their utility when examining issues such as climate change or biodiversity loss.	Example learning outcomes: 1. Appreciate that sustainability demands participation from all disciplines and contributions from society.	Example learning outcomes: 1. Engage in self-assessment, self-reflection, and analysis and have strong awareness of one's own values and how they inform one's ways of seeing (i.e., paradigmatic awareness).

(Continued)

Table 9.1 (Continued)

Student Sustainability Attributes			
Holistic systems thinking: Everything is connected	Sustainability knowledge: Understand the context, know the challenges	Awareness and integration: Connect what I know with what you know	Acting for positive change: Contribute to co-creating a better future
2. Understand how tipping points, interdependencies, feedback loops, and emergent properties impact a variety of social, economic, and ecological systems. 3. Engage in dynamic conversation about different types of systems and processes (e.g., the food web, globalization).	2. Demonstrate an understanding of how social equity contributes to global sustainability. 3. Display knowledge of how economic theory and resource equity contribute to sustainability.	2. Empathize with intercultural perspectives and recognize their value to illuminate environmental and social issues. 3. Demonstrate the ability to weigh multiple perspectives.	2. Use relevant theories of societal and institutional change to identify when and where to direct energy and actions towards a targeted outcome. 3. Work collaboratively with others to creatively solve a community-focused problem.

measure learning about "sustainability," including (1) holistic systems thinking, (2) sustainability knowledge, (3) awareness and integration, and (4) acting for positive change. However, wanting to remain open to emergent learning in the field, we take this set of attributes less as a static, immutable endpoint than as a starting point for discussion and critical interrogation, helping us to avoid glossing over the complex and variegated ways the concept is deployed in specific contexts.

On our field school, students are tasked with unpacking how "sustainability" has been understood, struggled over, implemented, and resisted on the ground in specific, complex contexts we visit. The course does not represent an uncritical celebration of "sustainability" but rather a critical interrogation into its deployment and the political work it attempts to accomplish. A rigid set of outcomes as well as some form of pre- and post-testing for definitional clarity is thus insufficient for our purposes. Rather, the field school's organizing concept of "sustainability" is open, evolving, and highly contested, as we believe are many other concepts approached in field programs – for example, globalization, colonialism, or intercultural dynamics. In a field school, one does not come to some final, singular understanding of any of these concepts (see Musisi, this volume), a fact that tests our creativity in assessing learning.

In this context, we use an evolving set of interrelated outcomes that are open and flexible but robust enough to provide a way of tracing the evolution of the students' learning. An overarching objective of our field school is to cultivate a kind of "critical optimism," leaning on Latour's (2004) call to transcend "critique" and Robbins's (2004) metaphor of the hatchet and the seed. Students are continuously encouraged to interrogate the dynamics of the devastating "dominant social paradigm" by "cutting and pruning away the stories, methods and policies that create pernicious social and environmental outcomes" with a hatchet. Yet, to avoid leaving them feeling like cynical, disempowered, and indignant spectators, the field school connects students with inspiring people and projects that are, in the words of Hawken (2009, 1), "confronting despair, power and incalculable odds to restore some semblance of grace, justice, and beauty in this world."[10] Critical optimism is thus positioned as a challenge to both debilitating pessimism and oblivious optimism about our present condition and future opportunities (elaborated in Owens, Sotoudehnia, and Erickson-McGee 2015). Sustainability is understood in this context as a "fuzzy concept" (Gunder 2006), not to be uncritically celebrated nor cynically dismissed.

We can assess learning in terms of the proficiency with which students are able to wield the axe of critical interrogation while nurturing the seed of creative, positive action. More precisely, our assessment is guided by three learning outcomes that we have refined (and continue to refine) over time: (1) *enhanced, critical questioning*; (2) *reflexive, relational thinking*, and (3) *action with traction.*

Through analysing reflective essays, field journals, and in-depth interviews, we assess the extent to which students are able to ask more qualified, capacious, or conditional questions piercing the who, what, where, when, why, and how of sustainable community development, understanding the subject matter not in self-evident terms, but as a dynamic, contested social construction. The direction of questioning might vary from trip to trip and student to student, but a common expectation is that students can ask more piercing questions, in turn demonstrating their grasp of the variegated, complex, political context(s) within which the term "sustainable" is deployed and struggled over. For instance, a student who embarked on the 2012 trip hoping to identify the technology that would save the planet, came to question the role of technology itself, the relevance of technological solutions in different contexts, and how prioritizing technology may compromise other responses. A student on the 2014 trip wanted to understand how economic, ecological, and social priorities could be aligned through sustainable development, but through the course came to see that these concepts often obfuscate as much as illuminate what sustainability looks like on the ground. A number of students on the 2016 trip came to ask questions about the role of race in sustainable community development, elements that had not been recognized as relevant in their initial reflections and prior scholarly engagements.

Second, we correlate learning with the extent and depth of the connections students are able to make between course materials (e.g., lectures, readings, course framework), field learning situations, the diverse and often competing perspectives of individual presenters we meet, and their own lives. We explicitly ask students to document such connections in their field journals and other assignments. Taking bearings from Palmer (2007), we reflect on our own abilities as teachers "to weave a complex web of connections among [ourselves], [our] subjects, and [our] students" (10). A course is deemed successful to the extent that students no longer hold the study of sustainable community development at arm's length, but recognize their own deep embeddedness in their communities and their role in a sustainable future. Further evidence of the impact shines through as students reflect on the stories with which they return of inspiring people and examples, place-based memories, life lessons, and emotional connections, which are now a part of them as lifelong learners.

Third, to assess field learning, we look for evidence of behavioural change and sustained, positive engagement after the course finishes. While we are mindful of student autonomy, it would be disingenuous to suggest that we are disinterested in what students do with their learning.[11] The field school is premised on the conviction that converging social and ecological crises demand committed action, in terms of lifestyle change and/or political-community engagement. It is challenging to precisely evaluate how, and how much, students do things (differently) after the field school, given the diversity of student situations and backgrounds. However, through our follow-up interviews and Facebook posts, we are keen to track both individual and collective actions with traction.

Conclusion

Pressures to confirm benefits to administrators, concerns about potential pernicious effects, and the desire to ensure optimum impact compel the intentional assessment of field schools. This chapter has documented challenges and pressures that, in our experience, can confound this endeavour. We have explored challenges in terms of building capacity, ensuring a sufficient and appropriate scope of concern, accessing reliable data, and developing a flexible approach to running and evaluating field schools in the context of the relatively open field and a dynamic, contested conceptual terrain. We have also highlighted some possibilities for addressing such challenges, based on our own experiences assessing, refining, and delivering field schools.

Given that sustainability and many other conceptual bases for field programs are protean, evolving, and contested processes, rigid means of identifying learning are untenable. We remain leery of "the drive for clarity, transparency, and specificity" in the employment of learning outcomes (Hussey and Smith 2003, 357) and, rather, encourage expanding our radius of concern, being flexible and experimental. To the extent that students are able to undertake more penetrating, critical analyses; recognize their embeddedness in a dizzying, complex web of interconnections; and emerge as more engaged citizens, we are more confident that valuable learning is taking place "out there."

NOTES

1 Off-campus programs go by a wide range of names. Our experience, which informs this chapter, is with geography field schools, short-term travel study programs. While we expect the commentary to have broader relevance, we want to position ourselves for the reader.

2 At the time of writing, Cameron Owens is an associate teaching professor and the director and lead instructor of the University of Victoria (UVic) Geography Sustainability Field School program, while Maral Sotoudehnia is a UVic Geography graduate student and was the teaching assistant for the 2014 field school. Maral was involved in the assessment of the 2014 field school, which led to a publication: Owens, Sotoudehnia, and Erickson-McGee (2015).

3 For example, Owens hosted panels and paper sessions on field school assessment at the Association of American Geographers Conferences in 2015, 2016, and 2017.

4 Our evaluation process has evolved over the past six years. Each of our five field schools has been assessed using student reflective essays and follow-up interviews at the end of the course. The 2013 field school involved a second round of follow-up interviews four to six months after the program, conducted by a student. In 2016, we hosted an event attended by approximately 25% of the students from the five field schools. Students completed a questionnaire and participated in a group reflective exercise captured on video. A total of 107 students have participated in our evaluation.

5 Even if time and material resources are available, instructors may be suspicious and resistant to institution-led assessment efforts, especially if these efforts are imagined in terms of having to prove that their own work and the university's investment in these courses is worthwhile, significant considerations not required of non-field courses.

6 The URL for our website containing past student blogs and reflective videos is http://fieldschools.geog.uvic.ca.

7 The factors raised here featured in responses to the following question, which was part of student interviews after our 2013 field school and questionnaires completed in 2015 by students from the previous four field schools: Are there any ways the field school could be improved to provide a more transformative experience?

8 We use the term "producing" instead of obtaining data. To collect data implies that data are autonomous entities out in the field awaiting the observation of the researcher, while producing data implies that some things become data when they are identified and interpreted by the researcher. The latter is a much more integrative process that recognizes the inevitable performative act of the researcher in shaping the research subject (see Charmaz 2000).

9 Tucker, Gullekson, and McCambridge (2011) canvas a wide range of tools including the following: Cross-Cultural Adaptability Inventory, Generalized Ethnocentrism Scale, Global Competency and Intercultural Sensitivity Index, Intercultural Communication Apprehension Scale, Intercultural Development Inventory, Intercultural Learning Outcomes, Intercultural Sensitivity Inventory, International Awareness and Activities Survey, Multicultural Personality Questionnaire, and Trait Emotional Intelligence Questionnaire.

10 The following quote underscores the tone of our field program: "When asked if I am pessimistic or optimistic about the future, my answer is always the same: If you look at the science about what is happening on earth and aren't pessimistic, you don't understand data. But if you meet

the people who are working to restore this earth and the lives of the poor, and you aren't optimistic, you haven't got a pulse. What I see everywhere in the world are ordinary people willing to confront despair, power, and incalculable odds in order to restore some semblance of grace, justice, and beauty to this world. The poet Adrienne Rich wrote, 'So much has been destroyed I have cast my lot with those who, age after age, perversely, with no extraordinary power, reconstitute the world.' There could be no better description. Humanity is coalescing. It is reconstituting the world, and the action is taking place in schoolrooms, farms, jungles, villages, campuses, companies, refugee camps, deserts, fisheries, and slums. You join a multitude of caring people" (Hawken 2009, 1).

11 It should be clear that we reject the desirability or even possibility of detached scholarship but rather concur with Derickson and Routledge (2015) that we channel the resources and privileges afforded us in academia to supporting community partners and advancing social causes and insurgencies.

REFERENCES

Baxter, Jamie, and John Eyles. 1997. "Evaluating Qualitative Research in Social Geography: Establishing 'Rigour' in Interview Analysis." *Transactions of the Institute of British Geographers* 22 (4): 505–25. https://doi.org/10.1111/j.0020-2754.1997.00505.x.

Brewer, Elizabeth, and Kiran Cunningham. 2009. *Integrating Study Abroad Into the Curriculum: Theory and Practice Across the Disciplines.* Sterling, VA: Stylus.

Broadfoot, Patricia, and Paul Black. 2004. "Redefining Assessment? The First Ten Years of *Assessment in Education.*" *Assessment in Education: Principles, Policy & Practice* 11 (1): 7–26. https://doi.org/10.1080/0969594042000208976.

Burns, Anne, and John S. Knox. 2011. "Classrooms as Complex Adaptive Systems: A Relational Model." *Electronic Journal for English as a Second Language* 15 (1): 1–25.

Charmaz, Kathy. 2000. "Constructivist and Objectivist Grounded Theory." In *Handbook of Qualitative Research,* 2nd ed., ed. Norman Denzin and Yvonna Lincoln, 509–36. Thousand Oaks, CA: Sage.

Derickson, Kate Driscoll, and Paul Routledge. 2015. "Resourcing Scholar-Activism: Collaboration, Transformation, and the Production of Knowledge." *Professional Geographer* 67 (1): 1–7. https://doi.org/10.1080/00330124.2014.883958.

Dewey, John. 1916. *Democracy and Education: An Introduction to the Philosophy of Education.* New York: Macmillan.

Ellsworth, Elizabeth Ann. 2005. *Places of Learning: Media, Architecture, and Pedagogy.* New York: Routledge.

Elwood, Sarah. 2004. "Experiential Learning, Spatial Practice, and Critical Urban Geographies." *Journal of Geography* 103 (2): 55–63. https://doi.org/10.1080/00221340408978576.

Emery, Charles, Tracy Kramer, and Robert Tian. 2003. "Return to Academic Standards: A Critique of Student Evaluations of Teaching Effectiveness." *Quality Assurance in Education* 11 (1): 37–46. https://doi.org/10.1108/09684880310462074.

Fuller, Ian, Sally Edmondson, Derek France, David Higgitt, and Ilkka Ratinen. 2006. "International Perspectives on the Effectiveness of Geography Fieldwork for Learning." *Journal of Geography in Higher Education* 30 (1): 89–101. https://doi.org/10.1080/03098260500499667.

Gibson, Robert, Susan Hassan, James Tansey, Graham Whitelaw, and Selma Hassan. 2005. *Sustainability Assessment: Criteria, Processes and Applications.* London: Earthscan.

Giddens, Anthony. 1987. *Social Theory and Modern Sociology.* Cambridge: Polity Press.

Glass, Michael. 2014. "Encouraging Reflexivity in Urban Geography Fieldwork: Study Abroad Experiences in Singapore and Malaysia." *Journal of Geography in Higher Education* 38 (1): 69–85. https://doi.org/10.1080/03098265.2013.836625.

Gray, Alex. 2015. "Op-Ed: If You Want to BREW, You Should Not Study Abroad." *The Concordian,* 15 Oct. Accessed 21 July 2016. http://theconcordian.org/2015/10/15/op-ed-if-you-want-to-brew-you-should-not-study-abroad.

Guinness, Patrick. 2012. "Research-Based Learning: Teaching Development through Fieldschools." *Journal of Geography in Higher Education* 36 (3): 329–39. https://doi.org/10.1080/03098265.2012.696188.

Gunder, Michael. 2006. "Sustainability – Planning's Saving Grace or Road to Perdition?" *Journal of Planning Education and Research* 26 (2): 208–21. https://doi.org/10.1177/0739456X06289359.

Hawken, Paul. 2009. Commencement address, delivered at the University of Portland, 3 May. Accessed 21 July 2016. https://media.up.edu/Media_Relations/keynote_graduation_2009/keynote_graduation_2009_medium.html.

Hern, Matt. 2016. *What a City Is For: Remaking the Politics of Displacement.* Cambridge, MA: MIT Press.

Hope, Max. 2009. "The Importance of Direct Experience: A Philosophical Defence of Fieldwork in Human Geography." *Journal of Geography in Higher Education* 33 (2): 169–82. https://doi.org/10.1080/03098260802276698.

Hughes, Annie. 2016. "Exploring Normative Whiteness: Ensuring Inclusive Pedagogic Practice in Undergraduate Fieldwork Teaching and Learning."

Journal of Geography in Higher Education 40 (3): 460–77. https://doi.org/10.10 80/03098265.2016.1155206.

Hussey, Trevor, and Patrick Smith. 2002. "The Trouble with Learning Outcomes." *Active Learning in Higher Education* 3 (3): 220–33. https://doi.org/10.1177/146 9787402003003003.

Hussey, Trevor, and Patrick Smith. 2003. "The Uses of Learning Outcomes." *Teaching in Higher Education* 8 (3): 357–68. https://doi.org/10.1080/13562510309399.

Hussey, Trevor, and Patrick Smith. 2008. "Learning Outcomes: A Conceptual Analysis." *Teaching in Higher Education* 13 (1): 107–15. https://doi.org/10.1080/ 13562510701794159.

Kirby, Sandra, Lorraine Greaves, and Colleen Reid. 2006. *Experience Research Social Change: Methods Beyond the Mainstream.* 2nd ed. Toronto: University of Toronto Press.

Latour, Bruno. 2004. "Why Has Critique Run Out of Steam? From Matters of Fact to Matters of Concern." *Critical Inquiry* 30 (2): 225–48. https://doi .org/10.1086/421123.

McMorran, Chris. 2015. "Between Fan Pilgrimage and Dark Tourism: Competing Agendas in Overseas Field Learning." *Journal of Geography in Higher Education* 39 (4): 568–83. https://doi.org/10.1080/03098265.2015.1084495.

Mezirow, Jack. 1978. "Perspective Transformation." *Adult Education Quarterly* 28 (2): 100–10. https://doi.org/10.1177/074171367802800202.

Mitussis, Darryn, and Jackie Sheehan. 2013. "Reflections on the Pedagogy of International Field-schools: Experiential Learning and Emotional Engagement." *Enhancing Learning in the Social Sciences* 5 (3): 41–54. https:// doi.org/10.11120/elss.2013.00013.

Nairn, Karen. 2005. "The Problems of Utilizing 'Direct Experience' in Geography Education." *Journal of Geography in Higher Education* 29 (2): 293–309. https://doi.org/10.1080/03098260500130635.

Owens, Cameron, Maral Sotoudehnia, and Paige Erickson-McGee. 2015. "Reflections on Teaching and Learning for Sustainability from the Cascadia Sustainability Field School." *Journal of Geography in Higher Education* 39 (3): 313–27. https://doi.org/10.1080/03098265.2015.1038701.

Palmer, Parker. 2007. *The Courage to Teach: Exploring the Inner Landscape of a Teacher's Life.* San Francisco, CA: Jossey-Bass.

Pekrun, Reinhard, and Lisa Linnenbrink-Garcia. 2014. *International Handbook of Emotions in Education.* New York: Routledge.

Pilkey, Orrin, and Linda Pilkey-Jarvis. 2009. *Useless Arithmetic: Why Environmental Scientists Can't Predict the Future.* New York: Columbia University Press.

Robbins, Paul. 2004. *Political Ecology: A Critical Introduction.* Malden, MA: Blackwell.

Sauer, Carl. 1956. "The Education of a Geographer." *Annals of the Association of American Geographers* 46 (3): 287–99. https://doi.org/10.1111/j.1467-8306.1956 .tb01510.x.

Shurmer-Smith, Louis, and Pamela Shurmer-Smith. 2002. "Field Observation: Looking at Paris." In *Doing Cultural Geography*, ed. Pamela Shurmer-Smith, 165–75. Thousand Oaks, CA: Sage.

Silverman, David. 2015. *Interpreting Qualitative Data*. 3rd ed. Thousand Oaks, CA: Sage.

Stenglein, Ferdinand, and Simon Mader. 2016. "Cycling Diaries: Moving Towards an Anarchist Field Trip Pedagogy." In *The Radicalization of Pedagogy: Anarchism, Geography, and the Spirit of Revolt*, ed. Simon Springer, Marcelo Lopes de Souza, and Richard White, 223–46. New York: Rowman & Littlefield.

Stone, Deborah. 2013. *Policy Paradox: The Art of Political Decision Making*. 3rd ed. New York: Norton.

Tiessen, Rebecca, and Robert Hulsh. 2014. *Globetrotting or Global Citizenship? Perils and Potential of International Experiential Learning*. Toronto: University of Toronto Press.

Tucker, Mary, Nicole Gullekson, and Jim McCambridge. 2011. "Assurance of Learning in Short-Term, Study Abroad Programs." *Research in Higher Education* 14: 1–11.

West, Charlotte. 2015. "Assessing Learning Outcomes for Education Abroad." *International Educator*, Nov + Dec. Accessed 2 Feb. 2017. https://www.nafsa .org/_/File/_/ie_novdec15_ea.pdf.

EMBRACING COMPLEXITIES ...

It's difficult not to romanticize the villages of N'wamitwa.

Earth, the colour of copper. Rondavels with thatched rooftops, painted a variety of white, blue, yellow – some decorated with traditional Tsonga patterns.

On our first day in N'wamitwa, we drive through Nkambako, Mugwazini, Msiphani, and the village of N'wamitwa proper. As schoolchildren walk home, they catch a glimpse of the taxi van full of Canadian visitors. Dozens of kids begin to chase this odd sight, screaming with laughter as we wave out the windows. Their excitement spills into our van. I look at the other field school students. All smiles. We gush over the children in their school uniforms and laugh when two little boys manage to grab hold of the van and jump on the back. They press their faces against the glass window, and, still giggling, they jump off. Exhausted.

It's difficult not to romanticize the villages of N'wamitwa. But it hasn't been easy to share my experience with others back home.

In the media, Africa is portrayed as this vast stretch of desert land, its people starving and suffering from drought after drought, from "third-world illnesses," from poverty and poor resources – an entire continent always on the brink of collapse, chaos, and warfare. Through this lens, Africa needs any help it can get. This contemporary form of colonial logic affords a group of Canadian university students the presumption of competence. This thought alone makes me cringe and brings on shame-induced nausea.

With these deeply ingrained images of Africa always a backdrop in my conversations at home, I try to describe my experience as best I can. I try to describe poverty. I try to explain the many common experiences or circumstances that are relatable to both Canadians and South Africans. I've seen immense beauty and resilience in South Africa, a deep sense of community throughout the villages, a work ethic that is unparalleled back home. I found a best friend in Basani Ngobeni – the talented young woman who guided us around rural Limpopo. She speaks seven languages.

It's difficult not to romanticize the villages of N'wamitwa.

Local crime has become a serious problem throughout the communities. We walk through the overgrown Hleketani Community Garden, the essential infrastructure stolen over the prior year. Even the strongest work ethic can only generate a meagre income when there are so few employment opportunities available.

In Joppie village, collecting water is an ongoing struggle. The government water delivery system is poorly managed, causing weeks'-long delays to get water to the villages furthest from urban centres. We spend a few nights in Joppie, hosted at the home of Mphephu and Daniel Mtsenga. In the early mornings, their eldest daughter, Vina, makes multiple trips to the water pump station, checking if the water has finally made its way to Joppie.

On our last afternoon at the Mtsengas', three grandmothers arrive to show us how they grind the maize to make mealie-meal. They fill a large wooden mortar with dried maize and pound it into a fine dust using heavy poles. Josephine and Rosina work together, synchronizing each thrust with one another. Beads of sweat trickle along their faces as they go on and on. This is hard work. We all laugh as we take turns with the tools, failing to match the women's rhythm and strength.

It's been 2 years since the field school. In the classroom, I find myself recalling these memories, relating them to my studies in different ways. I reject the standard images of harsh poverty and bare existence. But I also worry about romanticizing N'wamitwa, conveying the idea that resilient people can simply make do in spite of the structures that oppress them. I realize now that it is far more complicated than either sentiment, which is a very important lesson to learn.

Liah Formby
Participant, University of Victoria's Colonial
Legacies Field School in South Africa

10

Transformation in the Field: Short-Term Study Abroad and the Pursuit of Changes

MICHAEL R. GLASS

Action boy seen living under neon,
Struggle with a foreign tongue,
Red sails make him strong,
Action make him sail along.

– David Bowie, "Red Sails"[1]

Transformation in the Field

Who doesn't enjoy going overseas? A central theme in David Bowie's song "Red Sails" is the romantic perspective of the able-bodied male traveller who seeks adventure and eventually returns home as another person. The sense that international experiences will be transformative is also a central presumption driving the production and promotion of field courses. Field courses are popular with students and faculty alike because they relocate the learning experience to a place where pedagogical activities are suffused by a sense of the exotic and unfamiliar. "Transformation" is an appealing and evocative term, indicating a change in form or appearance from a prior condition; however, such terms can be problematic when it comes to understanding and assessing the outcomes of short-term international field courses for individual students or, indeed, for entire cohorts.

There are different ways that transformation and its synonyms are used to describe field courses. For instance, labelling field courses as "transformative" may imply a simple *branding strategy,* used for promotion in an increasingly crowded market of similar opportunities provided by departments, university study abroad offices, and third-party providers. While perhaps benign, creating false expectations for students can undermine the putative value of field courses if returning students cannot detect any profound changes in their perspective. Another possibility is that "transformation" is being proffered as a *learning outcome* for students, where the goal for faculty is to use the setting of a field course to stimulate some kind of metamorphosis, transmutation, or conversion in the participants of the field course. Depending on how transformation is understood, this goal might also have value as a motivating principle for curriculum development, but the parameters of transformation must be defined. Furthermore, using a foreign destination as a backdrop for some anticipated object lessons presents some notable sociocultural hazards. For instance, immersion in a foreign culture with different idioms, language, and landscapes may reinforce (rather than dispel) perceptions of the "other" (Vibert and Sadeghi-Yekta, this volume; Nairn 2005). A third perspective on transformation refers to the domain of pedagogical research known as *transformative learning theory* (Taylor 2008, 2009; Merriam 2008). This particular theory looks to construe meaning from experience (Clark and Wilson 1991; Mezirow 1997). It is not always clear whether field course coordinators are creating an explicit point of reference to this school of research, but understanding the tenets of transformative learning theory might enable field course faculty to better assess whether or not they will present transformation as a goal for participants.

This chapter problematizes the use of "transformation" as an operating trope for short-term international field courses (defined as those lasting less than four weeks) by examining transformation as brand, transformation as outcome, and transformation as theory. Elsewhere in this volume, Owens and Sotoudehnia argue that the value of field schools cannot be presumed and that faculty must be self-critical to understand their influence on the learning experience. The conceptual critique included in this chapter concurs with their discomfort about presuming outcomes and is based on earlier critiques of how limited critical reflexivity can be in short-term international field courses (Glass 2014, 2015b). Critical reflexivity is both a part of the transformative learning paradigm and a part of field instruction that prepares students for professional research, deepens their learning, and provokes reflection on how their own identity and positionality affect their encounters with field sites. Using a review of reflexivity and transformative learning from geography and education theory, I describe four challenges faculty must address if they wish to create transformative field experiences complementing the challenges raised by Owens and Sotoudehnia. I support these arguments with information drawn from my own short-term field schools at the

University of Pittsburgh. I conclude by reflecting on the validity of using "transformation" as a goal for short-term international field courses.

Transformative Tropes in Field Courses

Transformation as Branding

Twenty-first century higher education is characterized by the internationalization of learning opportunities (Walker 2010). Global exposure is frequently lauded as an influential learning objective for students in undergraduate programs. The new emphasis on internationalization in the United States and Canada is reflected in the rise of study abroad offices on college campuses (Brewer and Cunningham 2010), the new requirement for global competence credits in undergraduate degrees (Williams 2009), and the new opportunities available to students to gain direct international experience (Salisbury 2015). Student opportunities are varied, falling generally along the dimensions of independent or group travel and short or long-term programs.

The reasons for the rising visibility of study abroad opportunities are well researched. Part of this new emphasis is because a more globalized world requires global fluency arising from exposure to different cultures and contexts (Eckert et al. 2013). In addition, the comparative accessibility of international travel and communication makes the logistics of conducting study abroad programs easier than in earlier eras. Also, the sense that international experience is now another marker of employability for graduates is spurring demand among students for these opportunities. The institutional presumption that global experience matters, linked to the presumption by students that studying abroad will provide value added to the "student experience," means that institutions of higher learning are now promoting the strength of their study abroad programs as part of their overall strategies to attract and retain students.

Universities in the United States and internationally are now using the promise of study abroad as a way to attract students and enhance their college experience. Although fewer than 10% of undergraduate students in the United States and only about 3.1% of students in Canada at present study abroad (Johnson 2016), these proportions have grown considerably since 1990, and study abroad has become a more formal and professional aspect of universities and colleges (Brewer et al. 2015). As part of the marketing and branding for study abroad, transformation is commonly cited as a reason for undergraduate students to travel as part of their degree. For example, the fifteen colleges of the Atlantic Coast Conference all include dedicated study abroad offices that facilitate student engagement in university or third-party international study programs. Each study abroad office has a website, and fourteen of the fifteen websites feature either mission statements, "why study abroad" testimonials, or both. A content analysis of these websites found that two schools (North Carolina State and Boston College) use "transformative experiences" as part of the argument in favour of studying abroad. The other twelve schools did not list transformation directly, but used synonymous terms such as "new" or "changed perspectives," altered "[global or cultural] understanding," and "personal growth." These claims of a transformative experience were occasionally supported by brief anecdotes from students who had participated in earlier programs, and by statements assuming that transformation can occur whenever students leave the campus environment. Transformation through study abroad also occurs in institutional mission statements. For instance, Macquarie University in Australia encourages "a culture of transformative learning in a research-enriched environment" as a strategic objective, with the Geography Department developing field courses to support that goal (Lloyd et al. 2015, 494).

The key point here is that international field courses are a part of a broader institutional push towards the internationalization of university instruction. Universities consider study abroad opportunities an important selling point for prospective students. As part of the marketing prospectus for international opportunities, study abroad offices cite the prospect that students will enjoy transformative opportunities. Such rhetoric (targeting both students and parents) is framed around terms including "perspective," "experience," and "growth" and strongly implies that returning students will be positively affected by their time spent abroad. These impacts include increased global awareness and competence, a unique and marketable skill set, and the capacity to enrich their campus-based training.

It is certainly the case that positive impacts may occur for students who venture into the field, and it is equally certain that marketing pablum is often benign and easily discerned by critical consumers. Uncritically overstating the prospects for a changed perspective is, however, problematic. For instance, field courses are highly variable in terms of the time students spend on the ground, the number of sites they visit, the size and dynamics of the group, and the competence of the instructors. Also, Norris and Dwyer (2005) note that study abroad programs (including field courses) can assume very different models – from an "island" approach where students do not interact with the local context to a highly immersive option that is sometimes presumed to provide better learning outcomes. Given these sources of variation, an uncritical institutional branding of study abroad courses as inevitable gateways to personal transformation can set false expectations.

Transformation as Outcome

Individual faculty can use transformation to describe anticipated learning outcomes for their field course. Within academic geography, recent research on field courses that cite transformation as a learning outcome tends to use the transformative learning theory of Jack Mezirow (1997) as a conceptual guide. The common threads that link these courses are a sense that students need to have their perspectives about a given process or place altered or corrected in some way, and that time in the field is a necessary precondition for this change. While direct experience is arguably an important pedagogical tool for geography instruction, not to mention other disciplines, a key risk is that international settings for field courses are used as a reductive foil for the instructor's own perspectives (Glass 2015a; Nairn 2005; Hope 2009).

Within environmental geography, transformation is claimed to be a desirable learning outcome, based on the assumption that students need to conceive of the planet differently in order to effect positive environmental outcomes. Some studies create a very broad definition of transformation, arguing that geography is fundamentally transformative because of its generally participatory and critical character (Kirman 2003). Most scholars agree, however, that transformation is something that must be actively and thoughtfully integrated into the curriculum. For example, Haigh (2014) describes introducing "Gaia theory" into his classroom to problematize and decentre the standard perspective that humans are at the epicentre of the earth's environment. Given how radically different this theory can seem, Haigh (64) uses the precepts of transformative learning theory to frame his presentation of such "troublesome knowledge." Outside the classroom, Ikibunle-Johnson (1989) describes a direct link between participation and action, claiming "knowledge, awareness, attitude and perceptions of grassroots people can be mobilized and transformed through participatory environmental education to generate motivations and skills for effective environmental management" (14). This link between engagement and transformative action also guides Owens, Sotoudehnia, and Erickson-McGee (2015) in their sustainability field schools in Cascadia. They write that "field study can encourage transformative and social learning as students transform how they see and act in the world through their social networks and the social contexts they interact with" (314). Their field assignments implement tactics to enhance the transformative potential of their time abroad, including direct experiences with varied stakeholders that are followed by reflective exercises to provoke introspective evaluation by the students. While the authors are candid about some of the challenges with respect to discerning transformation in the students, they note that seeing a capacity for students to pose more challenging questions is one measure of a transformed learner.

The challenge of assessing transformative outcomes is raised in several papers on field courses. Simm and Marvell (2015) use Krathwohl's (2002) taxonomy on the affective domain to assess student transformation in their sense of urban places visited on their international field courses. Using assessment devices, including reflective journals and exercises that promote active learning, the authors conclude that transformative outcomes were not uniform across students: "transformative change can occur in different ways and at different times: for some it arises from a dramatic event, gradual, fast or slow or even subconscious" (Simm and Marvell 2015, 610). Wright and Hodge (2012) also use transformation as a learning objective in their field course on development in Northern Australia. Their course is similar to Simm and Marvell's in that it emphasizes the affective domain of learning and looks to moments of discomfort as situations with the potential to transform the student understanding of Aboriginal community experiences. In their chapter in this volume, Castleden, Kiley, Morgan, and Sylvestre use a combination of field schools and digital storytelling methods in an attempt to transform the perspectives of students towards Canadian Indigenous world views. Their students create brief digital stories during field courses, eventually sharing them with a mix of community and academic representatives. The faculty argue that it was important for students to transcend academic engagement into a more emotional and spiritual connection with their subjects in order to prompt transformation. Also, Castleden et al. are careful to explain that completing a single field school is not sufficient to consider a student "transformed": "completion of one field school on Indigenous perspectives of resource and environmental management is, by no means, 'enough' in terms of transforming our awareness and understanding ... it is indeed a continual process of critical engagement about Indigenous-settler relationships in Canada."

What seems most important in these examples is that transformation is considered a valid and effective learning outcome to include within field schools. Some of these courses – particularly in the development and environmental fields – seem to use transformation as a crusading tool that can shift a student's perspective on a given issue. In others, particularly that of Simm and Marvell (2015), transformation is used more to assess the depth of encounter that students have with a field course site. In all cases, faculty are careful to note that field schools will provide *opportunities* for transformation. The geographer Peter Gould (1999, 269) was fond of saying, "I can teach you, I can't *learn* you." Learning is an active process, and transformative components of a field course are not necessarily seized by any or all students.

Transformation as Learning

Transformative learning theory was developed by the late Jack Mezirow, a sociologist and adult education researcher. It is an approach in education theory that tries to explain how adults learn (Kitchenham 2008). In setting out the value of transformative learning theory, Mezirow (1997, 5) asserted that "we must learn to make our own interpretations rather than act on the purposes, beliefs, judgments, and feelings of others. Facilitating such understanding is the cardinal goal of adult education. Transformative learning develops autonomous thinking." Mezirow began working on transformative learning theory in the late-1970s (1978), and drew in part upon Jürgen Habermas's (1984) distinction between instruction-based learning (i.e., the lecture model of information delivery) and a communicative mode of learning (i.e., multiple learners building consensus about new information that is presented). For Mezirow, transformative learning theory occurs in ten phases, beginning with a disorienting dilemma that can provoke higher-order learning (see Table 10.1). According to the model, the student will transition from the disorienting event through a period of (internal) self-reflection and (shared) rational discourse, until eventually the learner is capable of integrating the lessons learned from the dilemma into a new approach to their life

Transformative learning theory complements other approaches to adult learning, including Kolb's (1984) emphasis on direct experience as a source of learning, Bloom et al.'s (1956) cognitive taxonomy, and Krathwohl's (2002) cognitive, affective, and kinaesthetic learning

Table 10.1 Stages and Phases of Transformation

Stages of Transformation	10 Phases of Transformation
Disorientation	1. A disorienting dilemma
Self-Reflection	2. A self-examination with feelings of guilt or shame
	3. A critical assessment of epistemic, sociocultural, or psychic assumptions
Rational Discourse	4. Recognition that one's discontent and the process of transformation are shared and that others have negotiated a similar change
	5. Exploration of options for new roles, relationships, and actions
Action	6. Planning a course of action
	7. Acquisition of knowledge and skills for implementing one's plan
	8. Provisional trying of new roles
	9. Building of competence and self-confidence in new roles and relationships
	10. A reintegration into one's life on the basis of conditions dictated by one's perspective

Modified from J. Mezirow (2000, 22).

domains. As Simm and Marvell (2015) point out, Mezirow adds to these the importance of educational "baggage" for explaining the learned behaviours and philosophies that may impede an adult learner's development of new skills or perspectives.

Transformative learning theory is subject to criticism, especially within its native discipline of adult education. For instance, Collard and Law (1989) note that Mezirow's use of Habermas renders transformative learning theory subject to similar critiques to those levelled at Habermas directly: namely, that transformative learning theory is silent on how to foster ideal learning conditions in social contexts where structural inequalities make learning difficult, that it places inordinate attention on the individual and psychological vectors of learning, and that it is generally silent on the need to link perspective transformation to radical, emancipatory praxis. In a related critique, Clark and Wilson (1991) argue that the ahistorical nature of Mezirow's model of learning and self-reflection strips out those structural vectors (such as gender or age) that influence the development of perspective (see also Tennant 1993).

Mezirow (1989, 1991) responded to these critiques with enthusiasm, welcoming the opportunity to clarify and advance a theory that he considered to be incomplete. In his response to Collard and Law, Mezirow (1989, 171) concedes to the point that ideal learning conditions are by definition impossible to attain. Consequently, the challenge for transformative learning theory is to recognize and foster ideal discourse as a goal and to create an encouraging learning environment. Similarly, Mezirow (1991) considers cultural and historical context to be embedded within the theory, hence dismissing claims to the contrary. He is also critical of Collard and Law's presumption that transformation hinges on collective social action, arguing that it is challenging to define what this means. He stresses that educators should not confuse facilitating transformative learning with dogmatic indoctrination, arguing "the educator can be a partisan but a partisan only in a commitment to fostering critical reflection and action; the what, when, and how of the action is a decision of the learner" (Mezirow 1989, 172).

As noted in the prior section, the principles of transformative learning theory are found in several international field courses in geography. Transformation is likely considered a valuable teaching outcome and pedagogical device because it evokes an enduring and beneficial consequence of time spent abroad. Other disciplines also use transformative learning theory in their foreign-based instruction. For instance, Intolubbe-Chmil, Spreen, and Swap (2012) describe a consortium of universities in the United States and southern Africa (ESAVANA) who use transformative learning in a four-week study abroad course that engages with transdisciplinary perspectives on the Limpopo River Basin. They note that study abroad courses are a useful context for transformative learning theory, because the shared experience of time in a field provokes cognitive dissonance and enables "the opportunity [for learners] to critically reflect on their assumptions and to engage in coursework and interactions that prompt complex understandings" (Intolubbe-Chmil, Spreen, and Swap 2012, 168).

A common methodology for courses that implement transformative learning theory is to require pre-trip questionnaires or surveys that query students on their understanding of a given issue or theme that the program faculty will attempt to disrupt or problematize during the time abroad. For instance, the ESAVANA program requires students to write papers that give their perspective on southern Africa, and then uses these papers to "unpack their baggage" (Intolubbe-Chmil, Spreen, and Swap 2012, 172) upon arrival at their field school sites. This method corresponds with the first of Mezirow's stages of transformative learning theory (see Table 10.1), where a disrupting incident is used as provocation for changing perspectives in the field (see Vibert and Sadeghi-Yekta, this volume). The standard approach to evaluating transformation is to conduct post-test interventions with the field school participants, and perhaps with community partners (see Peifer and Meyer-Lee, this volume). These interventions can include interviews that occur at some interval after re-entry to the domestic environment (Simm and Marvell 2015; Owens, Sotoudehnia, and Erickson-McGee 2015), exams with reflective components (Bell et al. 2016), and faculty-student dialogue (Perry, Stoner, and Tarrant 2012).

Transformation in Short-Term Study Abroad Courses

Pitt Urban Studies in Southeast Asia

I turn now to describe my own short-term study abroad course, and describe ways that I attempt to encourage self-reflection in myself and my students. Reflexivity is an important aspect of transformative learning theory, yet it is a time-intensive goal by itself: one that I find challenging to incorporate in a short-term field course (Glass 2015b). The field course to Southeast Asian cities was developed to provide third- and fourth-year urban studies students at the University of Pittsburgh the option to conduct international fieldwork with department faculty. A prerequisite writing-intensive seminar is predicated on the question: "How do you conduct urban geographic research 'at a distance'?" Singapore and Kuala Lumpur serve as the focus cities for all student research projects and classroom activities. Most participants enter the course with a presumably similar unfamiliarity with the region's sociopolitical context and historical development.

The subsequent study abroad course focuses on core themes related to urban development, globalization, and cultural identity. In 2012, five students spent three weeks in the cities of Penang and Kuala Lumpur (Malaysia) and Singapore. In 2014, six students travelled to the region and spent two weeks in Singapore and Kuala Lumpur. Because of logistical constraints, students travelled separately to the field course site, rather than entering as a group. This meant that if the "disorienting dilemma" of the sort anticipated by Mezirow was initial entry into this new urban context, it did not occur for all students at the same time. The 2014 course involved significant group activities, including projects that were designed in the weeks prior to departure from the United States. Once students arrived, they engaged in classroom sessions, neighbourhood walks recording sound and vision, and visits to sites including hawker centres, government agencies, and museums.

Evaluation

Based on my experiences with the 2012 course, I developed evaluative criteria for the 2014 course that would allow me to assess the development of reflexive praxis in my students. This assessment also provides me with data that I can use to modify future field course curricula. Six pre-departure sessions helped reduce the number of contact hours required between students and staff in the field, thus providing students more time to encounter the field sites independently. The six students on the field course were assigned into two teams, and each was tasked with developing a research project based on their shared interests. The team-based learning exercise is an important precursor to encouraging reflexive practices in the field, as Bolton (2010) emphasizes the importance of using peers to strengthen feedback on one another's field observations. Peer learning seems especially important in international field contexts where limited time or capacity to interact with local groups could diminish the capacity to validate observations with local experts.

The course included both individual and group assignments. Individual assignments focused on the field research journal and extending a research paper written in the university-based writing course. The journal rubric explained that it was intended to "provide a critical, reflexive accounting of your field course experience," and a detailed explanation of expectations was given in the syllabus. Group assignments focused on the student-designed research project. Different unassessed exercises provoked the students towards collective reflexive practice in the field; for instance, open-ended questions and collaborative instructions encouraged student groups to think about how their research positionality could affect the research they were intending to undertake. Most student assessment occurred during the field course, and students were contacted for interviews approximately six months after the course concluded.

From a personal perspective, I consider it important that faculty are as self-critical as we hope students to be during their time abroad. Decisions about which sites to include or to omit during the field course, which itineraries to prepare for walking tours, and how to fashion student work were all considered, based on my experiences on the 2012 field courses (see Glass 2014; also see the chapter by Vibert and

Sadeghi-Yekta in this volume). I endeavoured to read student journals as frequently as possible during the two-week trip, but as I describe below, the parameters of a short-term study abroad course make such evaluation difficult to achieve.

Maximizing the Transformative
Prospect of Short-Term Field Courses

This chapter has outlined three different ways that transformation is applied in study abroad courses: either as a branding strategy, as a projected learning outcome, or as an intentional pedagogical device based in Mezirow's transformative learning theory. As the risks in branding a course as "transformative" should be clear, I conclude by considering four challenges that faculty may face with embedding transformation into a short-term field course. Short-term field courses are useful instructional frameworks because a course of fewer than four weeks may be less expensive to operate, can be inserted outside the regular semester (during a spring break, for instance), and creates fewer personal or professional disruptions for students and faculty.

Despite the appeal of short-term field courses, it is little wonder that these limited ventures into the field are at odds with some of the intensive methods that the literature on transformative learning theory prefers. While some scholars remain positive about the capacity for short-term study abroad to "foster transformative learning environments where new experiences and perspectives may be developed" (Perry, Stoner, and Tarrant 2012, 682), such claims are not supported by extensive evidence. Several of the examples of short-term study abroad that claim to be transformative lack methodological interventions that can adequately assess whether a changed perspective is achieved in many (or any) learners. My own field courses are no exception to this evaluative lacuna, given that the courses have occurred too recently (within the past five years) to allow for sustained, long-term evaluation of learner outcomes. This is not to say that short-term field courses do not have merit, or that they cannot be transformative. My experiences in 2012 and 2014, however, revealed several outcomes that complicate the objective of transformation (Glass 2015b). In particular, I argue faculty must be intentional about the reasons for using transformation and think through the following issues: assessment, resistance, scheduling, and evaluation.

Assessment Criteria

The challenge of assessing transformation is similar in many ways to that of assessing reflexive practices during fieldwork (Glass 2015b). In particular, how can a student's changed mindset be measured meaningfully, let alone in comparison with his or her cohort? As with training students to be more reflexive subjects, exercises that seek to provoke a transformative experience should be calibrated thoughtfully, with grading rubrics that emphasize personal growth rather than comparative evaluation across the student group. A key issue is to consider how students can examine and explain their own perspective on their field experiences while remaining flexible enough to acknowledge that changing one's perspective will mean different things for different students. Accordingly, standard assignments that provoke students to change their opinions about a given topic frequently rely on qualitative methods such as the reflective journal or diary (McGuinness 2009; Gleaves, Walker, and Grey 2008), peer review and debriefing (Kaufman 2013), as well as member checks and triangulation to provide outlets for individual and collective reflexive practice (Cresswell 1998). While these approaches allow for students to consider the concept in their own ways, there is still a problem of deciding how prescriptive to make the assessment mechanism. For instance, Dummer et al. (2008) found that while including detailed guidelines for reflective journals helped alleviate student frustration with the openness of the exercise, staff found adding detail reduced the possibilities for originality and experimentation as students presumably worried about deviating from the assessment rubric. As an added prosaic concern, field courses with large numbers of students will challenge the capacity of faculty and staff to review reflexive exercises that address the shifting transformation in a student's perspective.

Student Resistance to Transformation

Students may be uncomfortable with the notion of writing about their research positionality because it is not actively taught in high school settings or because the expectations are not as clear-cut as for other types of assignments. Kember et al. (2008) offer a four-category taxonomy for assessing reflexive writing that can provide some guidance for staff and students. The taxonomy places "habitual action/non-reflection" as the most rudimentary, least significant reporting, with "understanding" and "reflection" indicating a progression towards "critical reflection" marked by "evidence of a change in perspective over a fundamental belief of the understanding of a key concept or phenomenon" (379). Despite the availability of guides and rubrics defining what practices faculty expect students to model, students may still feel resistant to exposing themselves in an assessed exercise. Kaufman (2013, 79) notes that this resistance is a reminder that reflexivity is not achieved automatically and that developing the willingness and capacity for students to think and write about their research positionality is an incremental process – particularly "in institutions where the student body is more uniform and less diverse."

Another form of resistance that can occur is an unintended consequence of the "island" model of study abroad delivery. In this model, students and faculty travel to a site and do not engage in sustained interactions with local communities or places. Enabling such interaction is a challenge in short-term trips where the time required for meaningful interactions is limited. Students may not face the disorienting dilemma associated with transformative learning, nor gain access to alternative perspectives that could shift their understanding of a concept or process.

Scheduled Transformation

Emphasizing transformation within a short-term field course curriculum is challenging for two further reasons. First, incorporating activities that attempt to provoke deep-seated and sustained changes to personally held beliefs into a new or existing module may be complicated, given the number of competing requirements that must be included (Kent, Gilbertson, and Hunt 1997; Panelli and Welch 2005). For instance, different site visits, travel criteria, and the need to provide "down time" all mean that finding time for meaningful, potentially transformative encounters might be easier said than done within a filled curriculum. As with reflexivity, it seems from transformative learning theory that transformation as a learning tool is most effective when taught and conducted in an intentional and sustained manner. This means that instructors need to be committed to embedding precepts of transformative learning theory at multiple stages of the degree program, let alone during the field course. Although challenging to implement, such an approach may enable the module to achieve the "deep learning" that current field course pedagogy emphasizes (Herrick 2010; Malam and Grundy-Warr 2011; Hovorka and Wolf 2009), while providing students with a mode of engagement that allows them to reflect and act on a more personal level about their field experiences (Park 2003). Second, introducing practices such as reflexivity that might affect transformative learning during short-term field courses is problematic given the time demands of a short-term field course for contact hours with faculty and the need to use the available time to introduce students to new field sites (Glass 2014). These time limits may leave insufficient time for students to reflect on their experiences in any meaningful way or for staff to grade student products and provide feedback on how to develop the linkages between experience, positionality, and course concepts. This problem is most likely to arise in traditional, "teacher-centred" courses that emphasize the transfer of knowledge from faculty to students. A possible solution to this problem is to adopt a more horizontal, "learner-developed" course that brings students into the process of constructing learning objectives and course goals (Hains and Smith 2012; Marvell, Simm, Schaaf, and Harper 2013).

Evaluating Transformation

The fourth challenge to transformative learning in short-term field courses is the question of how to assess the outcomes. In the afterglow of an exciting, intensive, and occasionally frustrating travel experience, it may be natural for students and faculty to claim that the experience was tremendous and transformative. This is partly because the investment of time and money in the course leads participants to justify their engagement in the course. It might also be because there are professional pressures to claim that the field course was worthwhile: students want to please their faculty by claiming the trip was valuable, and faculty want to please their administrators and funders to ensure the trip is offered in the future! In either case, what I am most cynical about is the yearbook model of evaluation: the circumstance where hagiographical inscriptions written in high school yearbooks in the waning days of the school year are taken as unbiased proof that these were, indeed, the best days of their lives. It is highly uncertain whether even the most challenging or moving encounters during a short-term field experience will carry any sincerely transformative impact beyond the initial weeks of re-entry from the field. Furthermore, does it matter? The critics of transformative learning are quick to note that there is value in direct experience. However, can (or should) the consequences of these experiences be evaluated six months or more after a short-term field course, if action for social change has not occurred?

Conclusion

And these children that you spit on
As they try to change their worlds
Are immune to your consultations
They're quite aware of what they're going through

– David Bowie, "Changes"[2]

As a final epigraph, I return to David Bowie. His song "Changes" is justly famous for describing the artist's capacity for reinvention and self-reflex-ive positioning outside normative social frames of the late 1960s. The lyrics also convey a sense of the agency that occurs in the process of adult education. To the extent that instructors can establish any reasonable expectations for their students, the process of learning in the field is undoubtedly communicative. The interactions between faculty, students, and stakeholders in the places visited during a study abroad program will affect the types of transformation that occur: *they're quite aware of what they're going through.* While faculty and universities find great utility in transformation to describe international field courses, it is less clear whether field courses can have the results that their proponents claim because quantitative assessment of students post-graduation tends to be very limited.

Before entering the field with expectations of transformation, it is important to consider what this entails. At worst, transformation is used as a marketing ploy to increase course enrolment. This dilutes the impactful nature of transformation as understood by transforma-tive learning theory. A similarly misguided use of transformation is the indoctrination model of learning that Mezirow critiques. Faculty should remain open to the notion that transformation need not follow a nar-rowly prescribed shift in perspectives. Perhaps students come to rebuff the idea that a given model of city design or life is valid, or believe that a theory of human-environment relations is untenable, despite its popularity. At its best, transformation unleashes critical thought in adult learners through their golden years. Through an open dialogue between faculty and students, transformative learning entails a critical and reflexive approach to knowledge creation and recreation. Getting there is a challenge, and all parties should understand that effecting meaningful transformation is a lengthy process, unlikely to be achieved in a short-term study abroad course.

NOTES

1 "Red Sails," words by David Bowie, music by David Bowie and Brian Eno.
Copyright (c) 1979 EMI Music Publishing Ltd., Tintoretto Music and
Universal Music Publishing MGB Ltd. Copyright Renewed.
All Rights on behalf of EMI Music Publishing Ltd. Administered by Sony/ATV
Music Publishing LLC, 424 Church Street, Suite 1200, Nashville, TN 37219.
All Rights on behalf of Tintoretto Music Administered by RZO Music.
All Rights on behalf of Universal Music Publishing MGB Ltd. in the U.S.
Administered by Universal Music-Careers.
International Copyright Secured All Rights Reserved.
Reprinted by Permission of Hal Leonard LLC.
2 "Changes," words and music by David Bowie. Copyright (c) 1971 EMI Music
Publishing Ltd., Tintoretto Music and Chrysalis Music Ltd. Copyright Renewed.
All Rights on behalf of EMI Music Publishing Ltd. Administered by Sony/ATV
Music Publishing LLC, 424 Church Street, Suite 1200, Nashville, TN 37219.
All Rights on behalf of Tintoretto Music Administered by RZO Music.
All Rights on behalf of Chrysalis Music Ltd. Administered by BMG Rights
Management (US) LLC.
International Copyright Secured. All Rights Reserved.
Reprinted by Permission of Hal Leonard LLC.

REFERENCES

Bell, Heather, Heather Gibson, Michael Tarrant, Lane Perry, III, and Lee
Stoner. 2016. "Transformational Learning through Study Abroad: US
Students' Reflections on Learning about Sustainability in the South Pacific."
Leisure Studies 35 (4): 389–405. https://doi.org/10.1080/02614367.2014
.962585.

Bloom, Benjamin S., Max D. Englehart, Edward J. Furst, Walker H. Hill, and
David R. Krathwohl. 1956. *Taxonomy of Educational Objectives: The Classification
of Educational Goals.* Handbook I: Cognitive Domain. White Plains, NY:
Longmans, Green.

Bolton, Gillie. 2010. *Reflective Practice: Writing and Professional Development.*
Thousand Oaks, CA: Sage.

Brewer, Elizabeth, and Kiran Cunningham. 2010. "Capacity Building for Study
Abroad Integration." In *Integrating Study Abroad into the Curriculum: Theory and
Practice Across the Disciplines*, ed. Elizabeth Brewer and Kiran Cunningham,
236–48. Sterling, VA: Stylus.

Brewer, Elizabeth, Rachel Shively, Nick Gozik, Dennis M. Doyle, and Victor Savicki. 2015. "Beyond the Study Abroad Industry." In *Assessing Study Abroad*, ed. Victor Savicki and Elizabeth Brewer, 33–56. Sterling, VA: Stylus.

Clark, M. Carolyn, and Arthur L. Wilson. 1991. "Context and Rationality in Mezirow's Theory of Transformational Learning." *Adult Education Quarterly* 41 (2): 75–91. https://doi.org/10.1177/0001848191041002002.

Collard, Susan, and Michael Law. 1989. "The Limits of Perspective Transformation: A Critique of Mezirow's Theory." *Adult Education Quarterly* 39 (2): 99–107. https://doi.org/10.1177/0001848189039002004.

Creswell, John W. 1998. *Qualitative Inquiry and Research Design.* Thousand Oaks, CA: Sage.

Dummer, Trevor J.B., Ian G. Cook, Sara L. Parker, Giles A. Barrett, and Andrew P. Hull. 2008. "Promoting and Assessing 'Deep Learning' in Geography Fieldwork: An Evaluation of Reflective Field Diaries." *Journal of Geography in Higher Education* 32 (3): 459–79. http://doi.org/10.1080/03098260701728484.

Eckert, James, Mushtaq Luqmani, Stephen Newell, Zahir Quraeshi, and Bret Wagner. 2013. "Developing Short-Term Study Abroad Programs: Achieving Successful International Student Experiences." *American Journal of Business Education* 6 (4): 439–58. http://www.doi.org/10.19030/ajbe.v6i4.7943.

Glass, Michael R. 2014. "Encouraging Reflexivity in Urban Geography Fieldwork: Study Abroad Experiences in Singapore and Malaysia." *Journal of Geography in Higher Education* 38 (1): 69–85. https://doi.org/10.1080/030982 65.2013.836625.

Glass, Michael R. 2015a. "International Geography Field Courses: Practices and Challenges." *Journal of Geography in Higher Education* 39 (4): 485–90. https://doi.org/10.1080/03098265.2015.1108044.

Glass, Michael R. 2015b. "Teaching Critical Reflexivity in Short-Term International Field Courses: Practices and Problems." *Journal of Geography in Higher Education* 39 (4): 554–67. https://doi.org/10.1080/03098265.2015.1084610.

Gleaves, Alan, Caroline Walker, and John Grey. 2008. "Using Digital and Paper Diaries for Assessment and Learning Purposes in Higher Education: A Case of Critical Reflection or Constrained Compliance?" *Assessment and Evaluation in Higher Education* 33 (3): 219–31. http://doi.org/10.1080/02602930701292761.

Gould, Peter. 1999. *Becoming a Geographer.* Syracuse, NY: Syracuse University Press.

Habermas, Jürgen. 1984. *Reason and the Rationalization of Society*, vol. 1, *The Theory of Communicative Action.* Trans. Thomas McCarthy. Boston: Beacon Press.

Haigh, Martin. 2014. "Gaia: 'Thinking like a Planet' as Transformative Learning." *Journal of Geography in Higher Education* 38 (1): 49–68. https://doi.org/10.1080/03098265.2012.763161.

Hains, Brian J., and Brittany Smith. 2012. "Student-Centered Course Design: Empowering Students to Become Self-Directed Learners." *Journal of*

Experiential Education 35(2): 357–74. http://doi.org/10.1177/105382591203500206.

Herrick, Clare. 2010. "Lost in the Field: Ensuring Student Learning in the 'Threatened' Geography Fieldtrip." *Area* 42 (1): 108–16. http://doi.org/10.1111/j.1475-4762.2009.00892.x.

Hope, Max. 2009. "The Importance of Direct Experience: A Philosophical Defence of Fieldwork in Human Geography." *Journal of Geography in Higher Education* 33 (2): 169–82. https://doi.org/10.1080/03098260802276698.

Hovorka, Alice. J., and Peter A. Wolf. 2009. "Activating the Classroom: Geographical Fieldwork as Pedagogical Practice." *Journal of Geography in Higher Education* 33 (1): 89–102. http://doi.org/10.1080/03098260802276383.

Ikibunle-Johnson, Victor. 1989. "Managing the Community's Environment: Grassroots Participation in Environmental Education." *Convergence (Toronto)* 22: 13–23.

Intolubbe-Chmil, Loren, Carol A. Spreen, and Robert J. Swap. 2012. "Transformative Learning: Participant Perspectives on International Experiential Education." *Journal of Research in International Education* 11 (2): 165–80. https://doi.org/10.1177/1475240912448041.

Johnson, Tim. 2016. "Why Do So Many Canadian Students Refuse to Study Abroad?" *University Affairs / Affaires universitaires*, 25 May. Accessed 2 Feb. 2017. http://www.universityaffairs.ca/features/feature-article/staying-home-study-abroad/.

Kaufman, Peter. 2013. "Scribo Ergo Cogito: Reflexivity through Writing." *Teaching Sociology* 41 (1): 70–81. http://doi.org/10.1177/0092055X12458679.

Kember, David, Jan Mckay, Kit Sinclair, and Frances Kam Yuet Wong. 2008. "A Four-Category Scheme for Coding and Assessing the Level of Reflection in Written Work." *Assessment and Evaluation in Higher Education* 33 (4): 369–79. http://doi.org/10.1080/02602930701293355.

Kent, Martin, David D. Gilbertson, and Chris O. Hunt. 1997. "Fieldwork in Geography Teaching: A Critical Review of the Literature and Approaches." *Journal of Geography in Higher Education.* http://doi.org/10.1080/03098269786183.

Kirman, Joseph M. 2003. "Transformative Geography: Ethics and Action in Elementary and Secondary Geography Education." *Journal of Geography* 102 (3): 93–8. https://doi.org/10.1080/00221340308978530.

Kitchenham, Andrew. 2008. "The Evolution of John Mezirow's Transformative Learning Theory." *Journal of Transformative Education* 6 (2): 104–23. https://doi.org/10.1177/1541344608322678.

Kolb, David. 1984. *Experiential Learning: Experience as the Source of Learning and Development.* Englewood Cliffs, NJ: Prentice-Hall.

Krathwohl, David R. 2002. "A Revision of Bloom's Taxonomy: An Overview." *Theory into Practice* 41 (4): 212–8. https://doi.org/10.1207/s15430421tip4104_2.

Lloyd, Kate, Richard Howitt, Rebecca Bilous, Lindie Clark, Robyn Dowling, Robert Fagan, Sara Fuller, Laura Hammersley, Donna Houston, Andrew

McGregor, et al. 2015. "Geographic Contributions to Institutional Curriculum Reform in Australia: The Challenge of Embedding Field Based Learning." *Journal of Geography in Higher Education* 39 (4): 491–503. https://doi.org/10.1080/03098265.2015.1103710.

Malam, Linda, and Carl Grundy-Warr. 2011. "Liberating Learning: Thinking Beyond 'the Grade' in Field-Based Approaches to Teaching. *New Zealand Geographer* 67 (3): 213–21. http://doi.org/10.1111/j.1745-7939.2011.01213.x.

Marvell, Alan, David Simm, Rebecca Schaaf, and Richard Harper. 2013. "Students as Scholars: Evaluating Student-Led Learning and Teaching during Fieldwork." *Journal of Geography in Higher Education* 37 (4): 547–66. http://doi.org/10.1080/03098265.2013.811638.

Merriam, Sharan B. 2008. "Adult Learning Theory for the Twenty-First Century." *New Directions for Adult and Continuing Education* 119 (Autumn): 93–8. https://doi.org/10.1002/ace.309.

McGuinness, Mark. 2009. "Putting Themselves in the Picture: Using Reflective Diaries in the Teaching of Feminist Geography. *Journal of Geography in Higher Education* 33 (3): 339–49. http://doi.org/10.1080/03098260902742425.

Mezirow, Jack. 1978. "Perspective Transformation." *Adult Education Quarterly* 28 (2): 100–10. https://doi.org/10.1177/074171367802800202.

Mezirow, Jack. 1989. "Transformation Theory and Social Action: A Response to Collard and Law." *Adult Education Quarterly* 39 (3): 169–75. https://doi.org/10.1177/0001848189039003005.

Mezirow, Jack. 1991. "Transformation Theory and Cultural Context: A Reply to Clark and Wilson." *Adult Education Quarterly* 41 (3): 188–92. https://doi.org/10.1177/0001848191041003004.

Mezirow, Jack. 1997. "Transformative Learning: Theory to Practice." *New Directions for Adult and Continuing Education* 74 (Summer): 5–12. https://doi.org/10.1002/ace.7401.

Mezirow, Jack. 2000. "Learning to Think Like an Adult: Core Concepts or Transformation Theory." In *Learning as Transformation: Critical Perspectives on a Theory in Progress*, ed. Jack Mezirow, and Associates, 3–33. San Francisco: Jossey-Bass.

Nairn, Karen. 2005. "The Problems of Utilizing 'Direct Experience' in Geography Education." *Journal of Geography in Higher Education* 29 (2): 293–309. https://doi.org/10.1080/03098260500130635.

Norris, Emily Mohajeri, and Mary M. Dwyer. 2005. "Testing Assumptions: The Impact of Two Study Abroad Program Models." *Frontiers: The Interdisciplinary Journal of Study Abroad* 11: 121–42.

Owens, Cameron, Maral Sotoudehnia, and Paige Erickson-McGee. 2015. "Reflections on Teaching and Learning for Sustainability from the Cascadia

Sustainability Field School." *Journal of Geography in Higher Education* 39 (3): 313–27. https://doi.org/10.1080/03098265.2015.1038701.

Panelli, Ruth, and Richard V. Welch. 2005. "Teaching Research through Field Studies: A Cumulative Opportunity for Teaching Methodology to Human Geography Undergraduates." *Journal of Geography in Higher Education.* http://doi.org/10.1080/03098260500130494.

Park, Chris. 2003. "Engaging Students in the Learning Process: The Learning Journal." *Journal of Geography in Higher Education* 27 (2): 183–99. http://doi.org/10.1080/0309826032000107496.

Perry, Lane, Lee Stoner, and Michael Tarrant. 2012. "More Than a Vacation: Short-Term Study Abroad as a Critically Reflective, Transformative Learning Experience." *Creative Education* 3 (5): 679–83. https://doi.org/10.4236/ce.2012.35101.

Salisbury, Mark. 2015. "How We Got to Where We Are (and Aren't) in Assessing Study Abroad Learning." In *Assessing Study Abroad,* ed. Victor Savicki and Elizabeth Brewer, 15–32. Sterling, VA: Stylus.

Simm, David, and Alan Marvell. 2015. "Gaining a 'Sense of Place': Students' Affective Experiences of Place Leading to Transformative Learning on International Fieldwork." *Journal of Geography in Higher Education* 39 (4): 595–616. https://doi.org/10.1080/03098265.2015.1084608.

Taylor, Edward W. 2008. "Transformative Learning Theory." *New Directions for Adult and Continuing Education* 119 (Autumn): 5–15. https://doi.org/10.1002/ace.301.

Taylor, Edward W. 2009. "Fostering Transformative Learning." In *Transformative Learning in Practice.: Insights from Community, Workplace and Higher Education,* ed. Jack Mezirow, Edward W. Taylor, and Associates, 3–17. San Francisco: Willey.

Tennant, Mark C. 1993. "Perspective Transformation and Adult Development." *Adult Education Quarterly* 44 (1): 34–42. https://doi.org/10.1177/0741713693044001003.

Walker, Patricia. 2010. "Guests and Hosts – the Global Market in International Higher Education: Reflections on the Japan-UK Axis in Study Abroad." *Journal of Research in International Education* 9 (2): 168–84. https://doi.org/10.1177/1475240910374228.

Williams, Tracy Rundstrom. 2009. "The Reflective Model of Intercultural Competency: A Multidimensional, Qualitative Approach to Study Abroad Assessment." *Frontiers: The Interdisciplinary Journal of Study Abroad* 18 (Fall): 289–306.

Wright, Sarah, and Paul Hodge. 2012. "To Be Transformed: Emotions in Cross-Cultural, Field-Based Learning in Northern Australia." *Journal of Geography in Higher Education* 36 (3): 355–68. https://doi.org/10.1080/03098265.2011.638708.

EDUCATION CAN BE EMPOWERING ...

Our first visit to Isha School initiated in me a new attitude and appreciation for quality education. When we arrived at the school, down a country road, it was beautiful. We were ushered through the open walls to the office. There were also some murals and beautiful plants, including palm trees, that lined the large field out back.

Then we went to the morning assembly and stood before hundreds of beautiful sun-kissed rows of children. They clapped for us and welcomed us. I soon realized that the most prominent contribution to the beauty I first observed was the overall feel and energy of the school. There was a definite sense of gratitude and focus, which I've never before witnessed to this degree in a school.

After the assembly, our class met the principal, whose presence commands respect. We saw how passionate she is about her work. I was surprised when she explained that the majority of these children whom I had just seen in their orange and tan uniforms live in extreme poverty. For many, the only meal they have each day is the hot lunch provided by the school. It became clear to me this is a large part of the reason for such gratitude. For some, this school is a matter of survival.

But this experience has given me the awareness that, if I choose, my education can be empowering. I not only know this with my mind, but also I have found a new gratitude and inspiration for school in my heart.

Emily Tennent
Participant, University of Victoria's
Applied Theatre Field School in India

Concluding Remarks

Out There Learning: Critical Reflections on Off-Campus Study Programs
describes aspects of pedagogy, place, and the potential for a meaningful
assessment of short-term off-campus study programs, often referred to as
"field schools" and "study tours." Four threads have been sewn through
each of these three sections of the book. The way the individual authors
decided to sew in the threads varies from chapter to chapter. When discussing short-term off-campus study programs as curated learning experiences, for example, some of the authors in this volume reflect on the
choices that instructors make in designing their field schools (Williams;
Vibert and Sadeghi-Yekta); others focus on the ability to assess the processes and outcomes of the given *curation* (Peifer and Meyer-Lee; Owens
and Sotoudehnia; Glass). In addition, some focus on specific assignments or course learning goals (Castledon, Daley, Morgan, and Sylvestre; Curran; Owens and Sotoudehnia). Almost all of the authors in this
volume discuss the theoretical models they draw from in the learning
and teaching processes of running an off-campus learning experience.
These consist of critical feminist pedagogies, including discussions of
positionality (McKinney; Williams; Musisi; Castledon et al.; Vibert and
Sadeghi-Yekta; Owens and Sotoudehnia; Glass) and the use of models
that attempt to decolonize the pedagogical experience or to highlight
Indigenous pedagogies, while also acknowledging the history and legacy
of racism and colonialism (Castledon et al.; Borrows; Curran; Vibert and
Sadeghi-Yekta; Owens and Sotoudehnia), as well as negotiations of transformative learning pedagogy (Musisi; Castledon et al.; Owens and Sotoudehnia; Glass), among others.

The second of the book's threads addresses cross-disciplinary and
interdisciplinary perspectives in short-term off-campus study programs.

By bringing together authors and editors with a wide range of disciplinary backgrounds – ranging from applied theatre, earth and ocean sciences, environmental and public health, Germanic studies, history, international education, law, psychology, and urban planning – this volume benefits from an understanding of specific disciplinary practices and traditions as well as from our own collaborative discussions about administering, teaching, and assessing field schools and study tours. These discussions made us aware of some of the institutional differences that exist at our respective colleges and universities as well as of the similarities and differences of our own academic experiences and the pedagogical choices we make. This also has to do with the way we view learning and the assessment of learning, a topic that we explore further below.

The third thread that is sewn throughout the book is the transformative potential of the off-campus learning opportunities under discussion. While many of the book's contributors, including those of the student-authored vignettes, have witnessed glimpses of the transformational potential of field schools, some of the authors in this volume question whether transformation can happen over such a short period of time (Owens and Sotoudehnia; Glass). Similarly, we elaborate on this topic further towards the end of this concluding chapter.

The fourth, and final, thread worked throughout this book looks at the complex relationships that emerge through community engagement. This topic is obviously most detailed and prevalent in the second section of this book, on place, but it is an important thread that is an essential part of the entire work. In the first section, on processes, it becomes clear that pedagogy and place are inseparable when leaving campus, and the third section, on assessment, offers ways of assessing and contemplating the benefits and drawbacks of certain types of community-engaged learning. The value and consequences of community interaction on short-term study abroad programs is an area in need of further study. While there is potential to form reciprocally positive relationships, there is also the possibility that the short-term visit can be harmful or, at least, unhelpful to the communities visited.

As we approached the completion of this book, the authors began to mull over how best to conclude this volume. We were emphatic that the discursive and reflective spirit that had guided us throughout would similarly guide the concluding chapter. After the authors submitted their final draft chapters, we convened a meeting of contributors at the University of Victoria.[1] Those unable to attend in person connected through Skype and Google hangouts, forming a rather comical, sci-fi-esque sharing circle of humans and laptops. In the dialogue, we reflected on the process of collaboratively writing this book, our thoughts on what teaching off-campus study programs has meant for us, and what it has revealed about teaching and learning "out there."

What follows is an edited and abridged version of this dialogue, which, we hope, reflects the collective spirit of this project and shares useful insights for others thinking about teaching and learning on short-term off-campus study programs. We have organized this conversation to emphasize three themes that emerged in the analysis of the transcript: (1) the variety of ways authors from different disciplines approach and value "out there" learning; (2) the different perspectives of authors on the ostensible binary between field and classroom learning, which includes reflexivity with respect to the positionality of the instructor; and (3) the authors' reflections on the notion of "transformation" as an overarching objective for field learning. Finally, we conclude with some brief thoughts about future research needs and directions. Although we could have continued our discussion for hours, we had to wrap it up at a certain point to attend to other meetings and obligations. Yet, even if the end of that live and "virtual" conversation was abrupt, the connections we made and the relationships we developed in the process of writing this book will allow further discussions to continue long after its publication. Through our collaborative endeavours, we built a sense of community that seeps through this concluding chapter, which demonstrates the desire to let our diverse voices be heard while addressing one another by our first names, which seems much more natural at this point in the writing process.

Complexity, Relationality, and Shared Vulnerability: Approaches to Out There Learning

In our concluding conversation, many participants note that one of the most interesting things about this project is how the various chapters reveal the diverse ways in which scholars, from a wide variety of disciplines, approach and value learning "out there." As Janelle Peifer observes, diverse disciplinary perspectives reveal that "there are so many different ways to think about and talk about these core issues," which can manifest very differently according to the perspectives and methods of the instructor and students. At the same time, we are intrigued by the common themes that emerged when scholars reflected on their programs. For instance, Kacy McKinney approaches her field school in rural India from a critical feminist perspective. She sees it encouraging students to learn and think relationally – to map connections rather than defaulting to difference, strangeness, or the process of "othering" when studying the world around us. Reading through their reflections five years after the program, she sees students coming to terms with complexity in the world – accepting that they do not know everything and gaining a sense of endless curiosity and excitement around the unexpected. Kacy sees students recognizing and embracing the sense of shared vulnerability, and building together a sense of intimacy, trust, and confidence that allows them to be comfortable with not knowing, with asking questions, and with making mistakes. Finally, the study abroad program helps students recognize their privilege and positionality. They were meant to be learning about sustainable development in India, but they came away being able to think more critically about where they stand in the world and the vast network of connections of which they are a part. Kacy would say, to herself and to the students, "We are here to learn about South Asia, but on a much bigger level I'm hoping to help you see how you can look at the world through a sense of connectedness, rather than looking first for the exotic and the different and the separate from you." The reflexive nature of much feminist pedagogy assisted with these processes.

Helga Thorson finds a similar place for affect and personal reflection in the Holocaust memorialization field school. During their three weeks in Europe (and even during the preliminary classroom week at UVic), students often bring together the rigorous scholarly analysis of academia with their emotional responses when meeting Holocaust survivors or visiting historical sites. This merging of scholarship and emotion is often unfamiliar to students, teaching assistants, and faculty alike. Several other instructors also emphasize the importance they place on students learning to reflect deeply on their positionality and to learn relationally in the field. For Matthew ("Gus") Gusul, a contributing editor of this volume and leader of an intergenerational theatre field school in southern India, an important part of interrelation is playfulness. Theatrical preparation and performances "provided a joyful, transformative experience [for participants]. I feel it is important to emphasize the joyful and fun element of learning out there." Meanwhile, for Nakanyike Musisi, a key challenge and aim is to bring such benefits of "learning out there" back into the classroom in Toronto – both to harness the energy and affective elements of field learning and to share those benefits with students back home.

The role of place in field-based learning is the focus of the evocative essays in the second section of this volume. Chapters by Deborah Curran and John Borrows emerge out of the discipline of law and are grounded in the notion that one cannot understand law, and especially Indigenous law, without understanding the place – the land, ecosystem, or socio-ecological context – that is the foundation of culture and law. Deborah's field course on the Central Coast of British Columbia "is nothing but contextualized, place-based learning" that examines the various disputes over the past fifty years in that particular location, from robust colonial and Indigenous legal perspectives. This nuanced emphasis on context makes things complex for students again; it helps to break down binaries between colonized and colonizer, Indigenous and colonial legal systems, "us" and "them." Students become much better able to take the legal principles and to deal with humans – "to look them in the eye and try and empathize and understand the context they're coming from," rather than just applying abstract legal principles to an issue or individual.

Reflecting on the chapters in Section 3, on assessing the value of field schools, several authors observe that key values of field-based learning lie in the experience of immersion and continuity, active engagement, learning in place from multiple perspectives, and building a reflective learning community. Cameron Owens looks for evidence of learning in students' emerging ability to question – to ask "more qualified, conditional questions" – and to identify complex interconnections. Ultimately, Cam hopes for change in behaviour, towards "action with traction." Michael Glass, whose programs explore themes of urbanization, globalization, and cultural identity in cities of Southeast Asia, reiterates the importance of immersion and peer-learning. At the same time, he expresses concern over time constraints and the limited capacity for interaction with local groups and communities. Both Cam and Michael interrogate the notion of "transformation," which Michael calls the "ostensible but undertheorized goal of many field programs." Both caution against automatically assuming that these programs are of benefit to students and the communities they visit, and they call for further investigation into how to comprehensively evaluate field-based learning programs.

Janelle Peifer and Elaine Meyer-Lee, who are responsible for institution-wide field school programs, focus on institutional-level change and how to incorporate macro variables into the assessment process to better understand transformation at the institutional level. Their concern is with assessment of processes – yet they keep a close eye on context in order to understand goals and outcomes in terms of socio-emotional and host community variables, as well as institutional aims. Recognizing the imperfections and limitations of assessment, Janelle and Elaine argue for trying anyway. They look at different definitions of transformation, and consider how people might think about them differently depending on where they're going, what they're teaching, and the reason behind the assessment.

Decentring the Field versus Classroom: Instructor Positionality and the Co-production of Knowledge

All the authors in this volume believe in the promise of "out there" learning, otherwise they would not invest the enormous time and effort in developing such programs. Yet tensions emerge in instructors' approaches. One such tension lies in how instructors see the distinction between the field and the classroom. Nakanyike's explicit aim is to explore the possibilities offered by off-campus programs for enhancing on-campus pedagogy, teaching, and learning. How can we transfer what we are learning in the field to the classroom, so that students in campus classrooms may become more engaged with the world around them? She points out, for instance, that the field provides constant evaluation: "In the field, on a daily basis, we are constantly getting feedback from the students and the community ... when a student cries, when students quarrel in the field, when students refuse to talk to each other." Nakanyike wonders how to transfer that awareness into the classroom, where the opportunities for evaluation and feedback tend to be less immediate.

Deborah reflects on the kind of learning that she believes can only happen when people interact outside of themselves and outside the classroom. In the context of law, she explains, "we pretend that we train people to go out and help other people after three years. You give them a bunch of tools, they graduate and they step out the door, and then they actually assist people in their lives or dealing with problems." In the process of several years of offering the Hakai field school, situated in Indigenous lands and communities, "it became entirely clear to me that we cannot or do not do that in the classroom." Classroom-based learning, where students may be shopping for shoes on one half of their screen and taking notes on the other, does not offer the same kind of opportunities for them to gain skills as problem solvers, or to process complex problems, or understand community relations. Part of the value of learning outside the classroom, Deb observes, is simply that people are learning all the time: they learn while sitting down with each other and having a meal, or talking with someone from the community, or tripping over a log, or seeing the whales out in the ocean. Thinking back on their law field school experience, students talked about a physical, embodied experience of learning. As Deborah notes, students learning in the field "don't have a choice. They are simply embedded and learning all the time, and they can't get away from it." Discomfort itself has learning value.

Some participants in the conversation express discomfort with what they see as a binary being drawn between the classroom and the field, both in the final discussion among authors and in certain chapters in this book. Elizabeth Vibert agrees that there are important distinctions to be drawn between "out there" and classroom learning, but she worries about the many students who are unable to take up the opportunity of going on a field school, often because of socio-economic background or family responsibilities. She warns that "if we set up that kind of binary between the much more intense, deep things that can happen in the field and the more limited things that can happen in the classroom, we're leaving out a whole bunch of people." Cam reflects that he had set up such a binary in an early version of his essay, and was not conscious of having done so until others read his draft. This conversation again highlights the value of collaboration across disciplines and perspectives.

Several participants point out, however, that field opportunities can be close at hand – in a local park or community, for instance, rather than halfway around the world. The point is to move beyond the walls of the physical classroom and, as Deborah puts it, interact with a community or landscape in a facilitated and curated learning environment. Cam recalls a course he co-taught with municipal officials in Victoria, British Columbia, Canada, to look at sustainable transportation planning. Students who had been on a previous international field school observed that the class had the same "feel" as travel study. The "walk-along" methodology described by Monica Degen and Gillian Rose (2012) calls on students to co-learn with locally based peers: students learn through the simple experience of talking and walking together, which presumably could be put into practice on campus (if not inside the classroom). Nakanyike holds to the possibility of bringing that "feel" to the classroom space itself, in part by creating opportunities for students to "behave a little bit differently," as they might in the field.

Despite the (contested) possibility of bringing some of the value of the field experience closer to home, several instructors observe that duration is part of the value of much off-campus learning. Janelle notes that the deepening of the student-instructor relationship that may happen when they spend considerable time together is an aspect of the change that many students experience.

For Kacy, whose field school is the longest of those featured in this book, duration is an important element: "I live with the students for three months. We're in kind of an isolated place. They don't get to go anywhere or have breaks; it's totally different [than the classroom]." Yet Kacy tries to bring that experience to her on-campus teaching. In a course on the geography of South Asia, for instance, she aims to teach a new way of seeing the world – "to actually see how you connect in ways that are negative and positive and everywhere in between." Allowing study abroad to inform Kacy's classroom teaching involves being "a little less subtle" about those political lessons than she might previously have been. Finally, Helga reminds us that it is not only instructors who bring back new lessons and methods from the field. We need to ask ourselves more often how students bring their experiences back to the classroom, enriching classrooms in often profound ways.

Evaluating "Transformative" Learning

In response to sometimes anxious concern over whether students are "transformed" by their field experiences, and the nature and evaluation of that transformation, Michael offers straightforward advice: "Chill out! Do the thing." Offer the field course and don't worry about this notion of "transformation" that gets bandied about. It is perhaps excessive or overly optimistic to expect that any field course must be the "very special episode" of a student's entire post-secondary educational experience. Michael reminds us that, in the first place, students often resist being told that they will have a certain kind of experience. Transformation is not necessarily a reasonable course objective. Nor is it possible to schedule "transformation" – a special moment within our curricula, very carefully curated in an attempt to provoke some kind of shift in the students. Finally, there's the problem of timing, especially in short-term field courses with limited student-instructor or student-community exposure. Time for discussion, reflection, and integration is generally needed before meaningful change can take place. That time is in short supply in a brief field course. We cannot predict where students' lifelong journeys of learning will take them. Field school experiences may contribute to those journeys, or they may not. As Michael emphasizes, "All we can do is provide one point of context – one point of contact – that they then take forward." He cautions that we should not oversell the field course as a transformative experience, not least because of the way such selling feeds into neoliberal categorizations of what we are meant to be doing in the university.

Critical analysis of the meaning of "transformation" and the possibility of assessment provoked distinct responses from Elaine and Janelle, who are developmental and clinical psychologists, respectively, by training and whose job, in part, is precisely to assess transformative learning. Their field is to study change and growth in individuals – not uncritically, of course – and as Elaine crosses the country presenting workshops on assessment, she finds faculty hungry to better understand what is and isn't happening in their classes, in order to better teach. All of this is inherently an evaluative process. Elaine points out that in the current political climate in the United States, many educators "are asking very serious questions about our role" in shaping critical consciousness of the world.

While bearing in mind important disciplinary differences in approaches to evaluation and assessment, Michael returns us to the question of time. He asserts that it is very difficult to evaluate "transformation" using standard course experience surveys or teaching evaluations with a small cohort of five to ten students who are asked to rate their experience within a few weeks of completing the course. This being the case, he queries the wisdom of including "transformation" as a learning outcome on course syllabi. Deborah brings us back to the practice of immersion, or what she calls "embeddedness," and the opportunity that provides for a different kind of learning. She prefers the term "deep learning" to transformation: "Deep learning is something that sticks with the students ... changing how it is that they view the world."

Another danger of setting students up to expect "transformation" is the disappointment or frustration that may arise if it does not happen. Helga recalls that before her 2016 field school, students about to make the journey had the chance to meet with former field school students. The former students said things like "this is transformational. It's going to change your lives." Students heard this refrain so often that some began to feel it as pressure. They worried that they themselves were to blame if the experience did not change their lives. Helga explained that "it really caused a crisis for some of them – feeling the pressure of 'I have to change because of this.'" Michael warns against "selling students a bill of goods" where they spend several thousand dollars for two weeks in the field with an instructor and expect some kind of personal or intellectual transformation: "I would prefer that they feel like as a consequence of the field course they'll not only get three credits towards their degree," he says, "but they'll also come away with some knowledge and perspective." He cautions circumspection about the reasons for going, noting that transformation is likely to be more incremental – and possibly painful: "That pain takes time. ... I can't remember in any of the field courses I've run that students have come across a burning bush in the desert or a great moment of profundity."

Even if burning bush moments do happen for some students, what comes next? Cam reminds us of a book exploring spiritual awakening in various religious traditions. In *After the Ecstasy, the Laundry* (Kornfield 2000), the author observes that after a profound experience that may seem transformative, many seekers struggle to integrate the new awareness into the day-to-day drudgery of their lives. Gus notes that students cannot be trained for burning bush moments. His aim is to train people to critically reflect on whatever experiences they have in the field, and he prefers the concept of "deep learning" to "transformation." Contributing editor Duncan Johannessen reminds us that change can also be in a negative direction – towards apathy or depression, for instance. Kacy raises a thought-provoking final point in this discussion. As a feminist critical scholar, she says, while she may not want to insist to students that she is teaching for "transformation," nevertheless "I know that my work is about that. I know that my work is about challenging students to think differently." Categorical labels are less important than the aim of guiding students to learn to think in new ways – which is surely a shared goal of most forays beyond the four walls of the classroom.

Future Research Agenda

What are the horizons of research on learning "out there"? As mentioned, our conversation ended before we could fully engage in this question. Certainly, new technologies will help define a changing landscape, as is well considered in works like France et al.'s *Enhancing Fieldwork Learning Using Mobile Technologies* (2015). Participants in this volume share the view that longitudinal assessment holds much promise. Michael calls for "slow scholarship," seeking to gain insight – five or even ten years after the fact – into the kinds of benefits students may have gained from field experiences. Such long-term assessment allows us to think meaningfully about how activities and experiences in the field "percolate and blossom in unexpected ways." Kacy's qualitative analysis of student responses five years after travel and Deborah's recent collection of such data indicate the potential of long-term reflections to offer new kinds of understanding of the meaning of these courses in students' lives. All the authors agree that the collaboration involved in preparing this volume was very productive. Thinking through the challenges and promise of off-campus learning with scholars from different disciplines, with very distinctive kinds of experience, has given us new tools and prepared the ground for building better programs.

NOTE

1 We were inspired by the 2014 collected volume edited by Catherine Etmanski, Budd L. Hall, and Teresa Dawson titled *Learning and Teaching Community-Based Research: Linking Pedagogy to Practice* (University of Toronto Press) that also used a group dialogue to build the concluding chapter.

REFERENCES

Degen, Monica Montserrat, and Gillian Rose. 2012. "The Sensory Experience of Urban Design: The Role of Walking and Perceptual Memory." *Urban Studies* 49 (15): 3271–87. https://doi.org/10.1177/0042098012440463.

Etmanski, Catherine, Budd L. Hall, and Teresa Dawson, eds. 2014. *Learning and Teaching Community-Based Research: Linking Pedagogy to Practice*. Toronto: University of Toronto Press.

France, Derek, W. Brian Whalley, Alice Mauchline, Victoria Powell, Katherine Welsh, Alex Lerczak, Julian Park, and Robert Bednarz. 2015. *Enhancing Fieldwork Learning Using Mobile Technologies*. New York: Springer. https://doi.org/10.1007/978-3-319-20967-8.

Kornfield, Jack. 2000. *After the Ecstasy, the Laundry: How the Heart Grows Wise on the Spiritual Path*. New York: Bantam.

Appendix
Field School Briefings

Heather Castleden – Field School: Indigenous Perspectives on the Environment and Health

Institution: Formerly Dalhousie University, School for Resource and Environmental Studies, Halifax, Nova Scotia, Canada; now Queen's University, Department of Geography and Planning, Kingston, Ontario

Location: Mi'kma'ki (Nova Scotia)

Length: Pre-course reading assignment, 3 days of on-campus learning, 10 to 12 days of travelling throughout Mi'kma'ki, and 3-day post-course workshop

Participants: 5 to 15 students, 1 instructor (+ 1 TA when >6 students), 15+ Indigenous Knowledge holders

Costs: $1375 (CAD) field school fee in 2018 + tuition; does not include flights from Kingston to Halifax; does include meals, travel in Mi'kma'ki, and accommodation (mainly tents at campsites)

Key Themes and Learning Objectives: Themes include (1) Indigenous world views on health, the land, and the meaning of "place" in the world; (2) history of Indigenous health in Canada; (3) Indigenous-settler relations in Canada; (4) social constructions and cross-cultural perceptions of the environment; (5) historical and modern treaties, key Supreme Court of Canada decisions affecting Indigenous rights, and the UN Declaration on the Rights of Indigenous Peoples; (6) Indigenous management of natural resources; (7) environmental/social justice and health equity; and (8) healthy lands and healthy futures.

Learning outcomes include (1) to demonstrate rich understanding of the ongoing colonial relationship of Indigenous peoples and settlers in Canada; (2) to actively and experientially engage with Indigenous peoples' ontologies, epistemologies, and methodologies in place; (3) to understand the importance of key legal decisions as well as historical and modern treaties involving Indigenous peoples in Canada in terms of links between Indigenous rights to resources and Indigenous health and well-being; (4) to develop critical, analytical, and reflexive knowledge on Indigenous perspectives on the environment and health; (5) to hone teamwork skills, including the ability to work under pressure, in cross-cultural contexts, and difficult conditions; and (6) to apply oral-visual and technical skills to knowledge translation in ways that reach beyond an academic audience.

Challenges Faced and Lessons Learned: In past offerings, students felt they needed more preparation before "entering the field," and so I have added pre-course readings with an annotated bibliography for an assignment as well as 3 days of in-class discussion before our departure. Cost when the field school was offered at Dalhousie University was never identified as an issue (it was approximately $100/day per student all-inclusive), but it has been difficult getting the field school approved at Queen's University because of different rules and regulations. At the time of writing, the course still has not been approved despite my being the recipient of a Principal's Dream Course grant to offset the costs in the first 2 years; it's been nearly a year of negotiations at the department and faculty level. The logistical challenges of moving between communities and the realities of fluctuating schedules make it stressful, and so recognizing that my time needs to be on coordinating, not cooking, helped alleviate some stress.

Deborah Curran – Field Course in Environmental Law and Sustainability

Institution: University of Victoria, Faculty of Law and School of Environmental Studies, Victoria, British Columbia, Canada

   Location: Hakai Institute, Calvert Island, BC

   Length: 2 weeks total + time to complete final research project after course finishes

   Participants: 10 to 16 students + instructor + teaching assistants

   Costs: Program fee $300 (CAD) in 2015; the Institute covered food and lodging; tuition and travel costs, including flights, not covered

Key Themes and Learning Objectives: As an explicitly interdisciplinary course, the purpose of the course is to examine the complexity of law and environmental governance in a specific geographical context, in this case the Great Bear Rainforest. This focus includes how both Indigenous and colonial law have shaped a specific geography and how that place has shaped law and policy viewed through the topics of (1) Aboriginal rights and title, (2) science in law, (3) land use and marine planning, and (4) the impact of energy systems on remote communities, in order to help students understand the overlapping jurisdictional and governance systems that shape the region.

Challenges Faced and Lessons Learned: Beyond the usual travel and logistical challenges that are inevitable with remote locations and group activities, the students benefited most from interacting with members and staff of the Wuikinuxv and Heiltsuk First Nations on whose traditional territory the field course occurred. Fostering respectful interactions and shaping the course to "give back" to those communities posed ongoing challenges given the short length of the field course, the remoteness of the communities that were accessible only by boat, and the relative inexperience of the students.

Michael Glass – Integrated Field Trip
to Singapore and Malaysia

Institution: University of Pittsburgh, Urban Studies Program, Pittsburgh, Pennsylvania, USA

Location: Singapore and Kuala Lumpur

Length: 2 to 3 weeks

Participants: 6 to 10 students

Costs: Students pay tuition; program otherwise supported through departmental, area studies, and university-wide grant

Key Themes and Learning Objectives: The main themes of the field program include globalization, neighbourhood analysis, and cultural identity. Students engage with selected research papers and then further enhance their understanding of urban theory with on-the-ground research in selected Southeast Asian cities.

Challenges Faced and Lessons Learned: A primary challenge involves providing a rich experience for the students in such a short-term field course and navigating cultural differences.

Matthew "Gus" Gusul – Intergenerational
Theatre for Development: International Field School
in Tamil Nadu and Pondicherry, India

Institution: University of Victoria, Department of Theatre, Victoria, British Columbia, Canada

Location: Rural Tamil Nadu, India

Length: One semester, 15 September to 8 December 2014

Participants: 11 senior undergraduate students + 1 instructor + 1 teaching assistant

Costs: $500 (CAD) in 2014 + cost of flight; to off-set costs, students raised funds for the seniors' village where we stayed and donated the money in exchange for room and board

Key Themes and Learning Objectives: The field school centred on the creation of intergenerational theatre performance between senior actors from a unique seniors' village that houses orphaned and destitute seniors, called Tamaraikulam Elders Village, and a junior high school that focused on educating rural Indian students, called the Isha Vidhya School, sponsored by the Isha Foundation. Students participated in three courses with three different sets of learning goals. The first course was a practicum study where they trained theatre artists in Theatre for Development and Intergenerational techniques, analysed each workshop and rehearsal of intergenerational theatre, and examined process and product of Intergenerational Theatre for Development. They also participated in a course called "Neo-Colonialism in India and Indian Theatre Course" in which they began to develop knowledge of India and the partner organizations and gained cultural and historical knowledge on Tamil Nadu, Pondicherry, and India. The third course was a pre-departure and re-entry course in which the students prepared for culture shock, defined personal and group identity mythology, deconstructed personal and Canadian identity, prepared for reverse culture shock and to re-enter their home community, and prepared for sharing their Indian experience.

Duncan Johannessen (Assistant Instructor) – Field Course in
Geological Fieldwork and Understanding the Local Tectonic
Setting and Regional Geology

Institution: University of Victoria, Faculty of Science, School of Earth and Ocean Sciences, Victoria, British Columbia, Canada

Location: Various locations on southern Vancouver Island, BC
Length: Up to 2 weeks
Participants: Up to 28 students + instructors + teaching assistants
Costs: Tuition + program fee of $600 (CAD) in 2017, which partially offsets food, accommodation, and transportation costs
Key Themes and Learning Objectives: Students visit various locations around southern Vancouver Island and the Gulf Islands learning to take field notes, interpret rock formations, compile data, and make maps and stratigraphical sections, which are also interpreted into a geological setting or paleo-environment. Students are also challenged to take each of these pieces and fit them into the broader picture of the regional geology of the area, which includes interpreting the tectonic processes that resulted in the current location of the rocks and landscape form. In groups, students engage in a multi-day mapping exercise, which they complete largely independently.

Challenges Faced and Lessons Learned: This is the first time most of these students will have worked on larger multi-day mapping exercises and wrestled with interpreting and extrapolating their observations into maps and cross sections. This work involves making informed judgments based on limited information, which can be very frustrating – especially for students of science who have largely been learning in an environment where there is a "right" answer in a textbook somewhere. A geological map is a hypothesis based on available data, and the uncertainties can be large. Thinking independently and developing original ideas is a fantastic but very significant challenge. The instructors' role is often to facilitate clear and logical thinking while trying to minimize frustration and stress as students face this challenge. The rigours of long days, physical exertion, and inclement weather add significantly to the challenge. However, the strength of the challenge also accentuates the satisfaction in completion and the group bonding that occurs with the shared adversity faced by the groups.

Kacy McKinney and John Keith Goyden – Environment and Development in the Indian Himalayas

Institution: University of Washington Seattle, Jackson School of International Studies, Seattle, Washington, USA
Location: Uttarakhand, India
Length: 10 weeks
Participants: 13 students + 2 instructors/co-directors (2011 and 2012)

Costs: $5050 to $5750 (USD) including tuition, room, and board for 10-week term, but not round-trip travel

Key Themes and Learning Objectives: The program combined the study of the political economy of Indian development through analyses of gender, labour, and the environment with internship or research projects related to rural development in collaboration with a local non-governmental organization. This study abroad program provided a range of opportunities to engage in conversation and exchange with members of local communities on these and other themes.

Challenges Faced and Lessons Learned: The length of the program both opened incredible opportunities and presented challenges. We had very high expectations for what was possible during the program, and we challenged students to engage fully every day in experiential learning, study, service, and exchange with program partners and local residents. Students needed time to reflect and absorb, and we were not always able to create sufficient time for these important components. Students were also differently prepared for living abroad in a remote location. We lived together 24/7 during the entire length of the program, which was labour-intensive, and sometimes challenging for both students and co-directors. Finally, only one of the 26 students had engaged in Hindi language study prior to participating in the program, which put pressure on the co-directors and English-speaking program partners, especially in the context of projects and internships.

Elaine Meyer-Lee – Global Journeys (a Required Four-Credit First-Year Course in Our Summit General Education Curriculum Focused on Global Learning and Leadership Development)

Institution: Agnes Scott College, Atlanta, Georgia, USA

Location: 12 to 14 options each year, which since 2015 have included Bolivia, Central Europe, Chile, Churchill (Canada), Croatia, Cuba, Dominican Republic, Galapagos Islands, Ghana, Guatemala, Iceland, Jamaica, Manchester (UK), Martinique, Morocco, Navajo Nation, New York City, Nicaragua, Northern Ireland, Panama, Puerto Rico, Toronto (Canada), and Trinidad

Length: About 8 days' immersion experience embedded in the middle of a semester-long course

Participants: Typically 17 to 21 students + instructor + co-leader + student leader

Costs: The cost of our Journeys program is covered through a special endowment draw that the Trustees authorized as a growth strategy. It has, in fact, already been quite effective in raising net revenues (enrolment, retention, etc.), so that the plan is for those additional revenues to cover the costs in the future. We are also fundraising to endow specific trips. This internal funding shows extraordinary commitment to off-campus global learning for all our students regardless of financial need.

Key Themes and Learning Objectives: A requirement for first-year students, this course introduces students to global structures, systems, and processes and connects these concepts to first-hand immersion experiences. Drawing on a variety of disciplines, interests, and expertise, it explores complex and interdependent relationships across the globe. Students examine a set of themes through common readings and activities: Identity (Self/Other/Culture), Imperialism/Colonialism/ Diaspora, Ethics of travel, and Globalization.

Learning Objectives
(1) Students will be able to identify and describe through at least two different examples how globalization relates to the particular section topic and analyse its impact on the Journeys destination.
(2) Using specific examples from their Journeys course and the immersion experience, students will be able to compare and contrast the impact global processes have on dominant and marginalized cultures.
(3) Students will be able to evaluate some of the historical, political, economic, scientific, and cultural forces that shape global processes and outline topics for future research and analysis.
(4) Students will develop their ability to engage across differences.
(5) Based on their interactions with and their learning from community members at the Journeys destination, students will critically reflect on their own values, ethics, and assumptions.

Challenges Faced and Lessons Learned
(1) Defining sharp, narrowly defined terms and learning outcomes at the outset is crucial, as faculty have many different views. Clear agreed-upon outcomes are also the bedrock of good assessment. Faculty development and incentives are therefore imperative.

(2) Required global study for *a whole very diverse class* at one time, while bold and exciting, creates many logistical challenges beyond the expected pedagogical ones. We therefore built some tasks (such as passport attainment) into the coursework, and are providing even more support for specific identity groups, such as LGBT-identified students, students of colour, students with disabilities, and Muslim students, through both co-curricular workshops and an in-class guest lecture.

(3) Assessing intercultural competence through one specific program, such as Journeys, is difficult to tease out from within a larger multidimensional initiative and requires some creativity. We are therefore experimenting with multiple qualitative methods to focus on the Journeys experience, as described above. These assessment methods are often quite labour-intensive.

Nakanyike Musisi – International Course Module Program: African Oral History

Institution: University of Toronto, Faculty of Arts and Science (History), Toronto, Canada
  Location: Kampala, Uganda
  Length: 10 days
  Participants: 15 students + instructor + on-site coordinator
  Costs: The University covered all the travel costs, and students covered their visa and immunization costs. Each student contributed $100 (CAD) towards the field cost. The rest was covered through an internal grant from the Faculty of Arts and Science and departmental support.
  Key Themes and Learning Objectives: This module is part of an existing on-campus course that takes students to Kampala, Uganda, to gain practical experience in the practices of oral history. The course uses a biographical approach to investigate the lives and times of elite African women in the twentieth century. Four African women's biographies are used as an entry point in understanding Africa's engagement with modernity. The readings and discussions are organized around three elite African women and one couple: (1) Wambui Otieno (Kenya), (2) Adelaide Cromwell (Sierra Leone), (3) Miria Matembe (Uganda), and (4) Walter and Albertina Sisulu (South Africa). Only Miria Matembe of Uganda is still alive. This international course module is organized around her biography and the historical moment in Uganda's history.

The module has four learning objectives: (1) Students gain first-hand critical understanding of the social, economic, and political forces that shaped the life and character of Miria Matembe and their implications for the study and writing of her biography. (2) Students explore and obtain critical understanding of personal and collective difficulties from an African feminist and anti-oppression viewpoint and their implications for praxis. (3) Students gain and apply knowledge, skills, and techniques as well as some analytical competence in doing oral history and interviewing in a fieldwork setting. (4) Students develop critical thinking skills and an inquiring interest in issues that historians deal with along with a commitment to the ethical principles of fieldwork. Above all, *learning about the self*: students deepen self-reflective knowledge/awareness of/about the social location of self and its implications for the practice of history and global citizenship.

Challenges Faced and Lessons Learned: It was a very intensive program, and there was hardly any time for students to recover from jet lag.

Nakaanyike Musisi – Summer Abroad Program: Conflict and Community in Africa

Institution: University of Toronto, Woodsworth College, Toronto, Canada
   Location: Kenya
   Length: 3 weeks
   Participants: 12 to 15 students + instructor + on-site coordinator
   Costs: Tuition and travel covered by students. Financial assistance is available to eligible University of Toronto students. Awards range from $500 to $4500 (CAD).
   Key Themes and Learning Objectives: Designed as an intensive inquiry into the causes, consequences, and, especially possible responses to conflict in Africa, the course introduces the complexities of conflict, peace, and development work in Africa. The course has the following four objectives: (1) to equip students with a knowledge and understanding of key issues and theories relevant to the study of conflict and peacebuilding in Africa; (2) to avail the students of an opportunity to apply this knowledge to practical cases and issues; (3) to sharpen students' analytical reading, writing, and presentation skills; and (4) most importantly, to develop in the students an appetite to engage in the world around them.
   Challenges Faced and Lessons Learned: The biggest challenge was the high cost of the program. This meant that only a select few, relatively well-off students could afford such a course. Another challenge was

to keep students focused all the time, so as not to turn the experience into a tourist trip. It was also an effort at times to challenge some of the students' engrained prejudices against Africa, without silencing them. As a woman from the region, and also because of my age, some of the students tended to depend on me more than I think would have been otherwise the case; the mothering role was at times very draining for me.

Cameron Owens – Cascadia Sustainability Field School and Northern Europe Sustainability Field School

Institution: University of Victoria, Department of Geography, Victoria, British Columbia, Canada

Location: On-campus in Canada and multiple itineraries in the American Pacific Northwest (Cascadia) or Northern Europe

Length: 1 week preparatory work in Victoria, followed by 4 weeks of travel study, with a follow-up semester back in Victoria undertaking community legacy projects

Participants: 15 to 28 students + instructor + graduate student teaching assistant

Costs: $2900 (CAD) program fee in 2017 + tuition; does not include flights, meals, and entertainment costs

Key Themes and Learning Objectives: Students travel to selected cities to connect with community leaders, urban planners, academics, and other engaged citizens learning about the challenges and opportunities for building sustainable and inclusive cities. The tone of the field school is critically optimistic, inviting students simultaneously to uncover unjust and dysfunctional urban social dynamics while encountering hopeful and inspiring people and projects addressing socio-ecological issues. Along with developing certain field-specific and community life skills, students learn to ask more piercing questions about urban development trajectories, recognize complex political contexts and webs of socio-ecological relationships, and see more clearly their own intimate connections to processes of (un)sustainability. The intention is that students return with a repertoire of inspiring stories, memories, experiences, and the inspiration to take positive action with traction in their own lives and communities.

Challenges Faced and Lessons Learned: The field program is expensive, and while students have access to funding opportunities, it remains exclusive. To the extent that part of our subject is looking at inequalities in urban development, it is important for us to reflect on our own privilege. The experience of travelling, studying, and living together for a month can be intense and disorienting for students (and instructors). We have found it crucial to undertake extensive emotional preparation in our pre-travel week, to encourage self-care throughout the program, and to design the itinerary to afford personal time for students. Finding a productive balance between structure/guidance and space for student self-direction is a key challenge. Students have noted that our own little community, in which issues and conflicting perspectives arise, can be imagined as a microcosm to help us understand the broader community social issues we are studying. Still, the program's emphasis on complex issues and multi-perspectival analysis has proven frustrating to some students seeking simple, definitive answers. One response has been to encourage continuous group reflection through discussion "sharing circles," where students can check in with each other.

Helga Thorson – I-witness Field School on Holocaust Memorialization

Institution: University of Victoria, Department of Germanic and Slavic Studies, Victoria, British Columbia, Canada

Location: On campus in Canada and in three or four different countries in Europe

Length: 1 week classroom work in Canada; 3 weeks in Europe; 10 subsequent weeks to complete written work and public presentations (May–August)

Participants: 15 to 17 students + instructor + experiential learning coordinator

Costs: Actual cost per student is around $3600 (CAD) in 2018; students pay $2950 program fee including travel and accommodations within Europe, not including international flights or tuition; the remaining program costs are funded by community donors

Key Themes and Learning Objectives: The I-witness Field School is designed to explore ways in which the Holocaust is memorialized at various Central European historical sites, museums, and monuments and provides students with an opportunity to acquire a deeper understanding of antisemitism, racism, religious and ethnic intolerance, homophobia, and the stigmatization of mentally and physically disabled communities. By visiting sites in three to four different countries in Europe (most recently in Germany, Austria, Hungary, and Poland), students learn not only about the past in each of these locations, but also about the multiple ways that the history of the Holocaust is told in present-day contexts. Participants of this program engage in cross-cultural dialogues about the Holocaust and its memorialization with European students in several locations.

Challenges Faced and Lessons Learned: Like most short-term study abroad programs, costs are often prohibitive for students. The field school advertises the program costs at an amount significantly lower than the actual costs and attempts to make up the difference through fundraising. We are now trying to build an endowment to help cover some of the program costs so that we do not have to scramble to make up the difference each time the program is offered. Another challenge is figuring out the extent to which the order of the itinerary makes a difference in the students' overall experience or in the narratives we build as a group about Holocaust memorialization.

Elizabeth Vibert – Colonial Legacies Field School in South Africa

Institution: University of Victoria, Department of History, Victoria, British Columbia, Canada

Location: On-campus in Canada and in rural and urban South Africa

Length: 1 week classroom work in Canada; 3 weeks in South Africa; 10 subsequent weeks to complete written work and public presentations (May–August)

Participants: 12 to 15 students + instructor + doctoral student assistant

Costs: Program fee including all South African internal travel and costs, not including international flights or tuition: $2400 (CAD) in 2014

Key Themes and Learning Objectives: The Colonial Legacies Field School is designed to enable students to identify and explore the ongoing impacts and influences of South Africa's multiple colonial histories. Through community visits, workshops, volunteer activities with local civil society groups, museum and site visits, and oral history interviews, participants consider the impacts of colonial histories in everyday life and on rural and urban landscapes. They engage with projects seeking sustainable rural development and with grassroots anti-poverty initiatives; reflect on the history of apartheid and reconciliation; discuss community responses to HIV/AIDS; and reflect critically on such challenges as gender and development; land, labour, and the global economy; and modes of historical memory. Many students observe that the experience of learning deeply about colonialism in the Global South helps them to better understand colonialism's ongoing impacts at home.

Challenges Faced and Lessons Learned: The high cost of delivering an international field school makes the experience unfortunately exclusive (all UVic students received at least $500 in institutional travel support and some received up to $2000; non-UVic students were unable to access these funds). Many students fell ill for a few days from generic anti-malarial medication purchased in South Africa. The 35-hour classroom week just before departure meant some students started the trip already depleted; allowing a few days between classroom work and departure is advised.

Aaron Williams – Sustainability and Environment Themed Field Studies in Europe and Asia

Institution: University of Calgary, Department of Geography, Calgary, Alberta, Canada
   Location: Various itineraries throughout Europe and Asia
   Length: 4 to 6 weeks
   Pasrticipants: 18 to 40 undergraduate students
   Costs: All expenses, $7000 to $10,000 (CAD)
   Key Themes and Learning Objectives: Since 2002, I have been involved with twenty-six travel study programs focusing on sustainability from the perspective of urban, environmental, cultural, and historical geography in Europe. I have also been involved with sixteen trips exploring sustainability challenges and opportunities in Asia. In Europe, our itineraries focus on cities known for innovation in sustainability, livability, environmental management, and social equity. Specific themes include regional development (past and present), transit, sustainable urban design, environmental problems and policies, and conservation. Many of these foci are also relevant on our trips to Asia, where we also focus on the rapid social and environmental changes from hyper-globalization, as well as my personal research focus on rebuilding after the 2004 Southeast Asian tsunami. Both trips involve meeting with local experts where students learn about multiple perspectives on sustainability challenges, develop critical thinking skills, and learn to interpret and identify the web of interconnections between human and natural systems.
   Challenges Faced and Lessons Learned: A major challenge is to continue to evolve and adapt to the changing needs and learning styles of students. This includes the incorporation of new technologies and learning tools that allow the current generation of students to relate to

the environments of focus in the way they choose to communicate and understand their world. This has changed since I started so I, too, have had to adapt. Another challenge is to get both students and faculty to adjust to experiential learning as a different way of learning from that which typically takes place in on-campus classrooms.

Contributors and Editors

John Borrows holds the Canada Research Chair in Indigenous law at the University of Victoria, British Columbia, in the Faculty of Law. He is a member of the Chippewas of the Nawash First Nation at Neyaashiinigmiing on the shores of Georgian Bay in Ontario, and he spends his time searching for innovative ways to introduce students and colleagues to the beauty that characterizes this small corner of the world.

Heather Castleden holds the Canada Research Chair in Reconciling Relations for Health, Environments, and Communities at Queen's University in Kingston, Ontario, where she is jointly appointed to the Department of Geography and Planning and the Department of Public Health Sciences. In her research and teaching, she strives to bring Indigenous and Western knowledge holders – and their knowledge systems and methodologies – together to address issues of health, social, and environmental justice.

Deborah Curran is an associate professor at the University of Victoria, British Columbia, where she is jointly appointed to the Faculty of Law and the Faculty of Social Sciences, School of Environmental Studies. As the executive director of the Environmental Law Centre and having always called the west coast home, her research engages the subjects of water and regional or watershed sustainability. When not in the classroom or in front of her computer she is gambolling through the forest or on the sea with her family.

Kiley Daley is a PhD candidate with the Centre for Water Resources Studies at Dalhousie University, Halifax, Nova Scotia. His research focuses on environmental and public health issues in rural and Indigenous communities.

Michael R. Glass teaches at the University of Pittsburgh. Among his courses are international field courses and local community engagement courses. His research includes work on housing, urban imaginaries, and geographical theory. At last count, he has hidden twelve David Bowie song titles in his chapter in this book.

Matthew "Gus" Gusul is an artist, educator, activist, storyteller, development worker, and community organizer who has led community-based projects with a diversity of organizations throughout Western Canada, Latin America, and India. His most recent work focuses on the creation of community-based theatre performances in Tamil Nadu, India, with HelpAge India and delivering culturally appropriate and playful literacy projects in Indigenous communities throughout Alberta with Frontier College.

Megan Harvey recently completed her PhD in history at the University of Victoria, British Columbia. Her research focuses on Indigenous-state relations, the history of land claims in BC, and how story has been central to the dispossession and attempted repossession of Indigenous lands. She had the very good fortune to be a student in the Ethnohistory Field School with the Stó:lō and co-director (with Elizabeth Vibert) of the Colonial Legacies South Africa Field School.

Duncan Johannessen is a senior lab instructor in the School of Earth and Ocean Sciences at the University of Victoria, British Columbia. He is involved in the organization and instruction of multiple short-term local field excursions as well as two longer field schools in various locations in BC, the Rocky Mountains, and as far away as Cyprus. His main interest is in determining what circumstances contribute to making a field course become a positive transformative experience for students.

Kacy McKinney is an interdisciplinary feminist educator and scholar trained in critical human geography, and she is also a visual artist. She is an adjunct assistant professor in Food Systems and Society at Oregon Health & Sciences University in the Pacific Northwest of the United States. Her interests range from the uses of comics and graphic novels in social science research, to the politics of genetically modified organisms, and the study of children and young people working in agriculture. She is committed to social justice and to inclusivity in higher education.

Elaine Meyer-Lee is associate vice president of Global Learning and Leadership Development at Agnes Scott College in Atlanta, Georgia. She has taught, led, and assessed intercultural higher education for over seventeen years. Initially motivated by the impact of her studies abroad in France and Haiti, she earned a doctorate in human development and psychology from Harvard University. She is also currently president of NAFSA: Association of International Educators. She loves backpacking with her teenaged sons and eating good food.

Nakanyike B. Musisi is an associate professor at the University of Toronto. She is a former director of Makerere Institute of Social Research at Makerere University, Kampala. Her more than five years' experience with field schools has involved taking students to Kenya and Uganda. Her research interests are in gender, colonialism, missionary work in Uganda, social change, and education.

Cameron Owens is an associate teaching professor in the Department of Geography at the University of Victoria in British Columbia. He has been involved with international field schools since 2001, and he currently leads programs actively and critically engaging with sustainable community development in the Pacific Northwest (of North America) and the Atlantic Northwest (of Europe).

Janelle S. Peifer is an assistant professor of psychology at Agnes Scott College in Atlanta, Georgia. Agnes Scott is an incredibly diverse small liberal arts college for women in the US Deep South. She is passionate about exploring the ways that people connect across differences – both within and outside of their home cultures. In her research, clinical work, and teaching, she examines how deeply complex, intersectional identities inform how individuals experience the world around themselves. She enjoys dancing and lives with the love of her life, parents, and two children in Decatur, Georgia.

Kirsten Sadeghi-Yekta is an assistant professor in theatre at the University of Victoria, British Columbia. As a theatre practitioner, she has been involved in projects with different communities and in a variety of countries: working with the Hul'q'umi'num' community on Vancouver

Island, children in the Downtown Eastside in Vancouver, young people in Brazilian favelas, disabled young women in rural areas of Cambodia, adolescents in Nicaragua, and students with special needs in schools in The Netherlands.

Vanessa Sloan Morgan is a PhD candidate in the Department of Geography and Planning at Queen's University in Kingston, Ontario, and a project coordinator at the Health, Environment, and Communities Research Lab. She approaches teaching and learning as a potentially transformative act from their position as a queer white settler to unceded Coast Salish territories / Victoria, British Columbia, Canada.

Maral Sotoudehnia is a PhD Candidate in the Department of Geography at the University of Victoria in British Columbia. Her research interests exist at the intersections of critical data studies, more-than-human economies, feminist methodologies, and urban studies. Her doctoral research investigates posthuman figurations of capital in cryptocurrency (e.g., Bitcoin and Ethereum) communities and decentralized autonomous organizations (DAOs).

Paul Sylvestre is a PhD candidate in the Department of Geography and Planning at Queen's University in Kingston, Ontario. His research concerns the intersection of white settler affect and practices of white settler possession and how these crystallize through the dominant spatial processes that produce settler urban space.

Helga Thorson is chair of the Department of Germanic and Slavic Studies at the University of Victoria in British Columbia. She directs the I-witness Field School on Holocaust memorialization in Germany, Poland, Austria, and Hungary. Through a grant from UVic's Learning and Teaching Centre she began the Field School Forum series, bringing together faculty and staff at UVic involved in running field schools. These sessions provided the impetus to embark on this book project.

Elizabeth Vibert is an associate professor of history at the University of Victoria, British Columbia. An historian of poverty and colonialism, since 2012 she has been privileged to be involved in community-engaged research with older women in a rural community in South Africa. This research has led in unanticipated directions: she launched the documentary film *The Thinking Garden* in 2017 (dir. Christine Welsh), and she founded and directs UVic's Colonial Legacies Field School in South Africa.

Aaron Williams is a senior instructor in the Department of Geography at the University of Calgary in Calgary, Alberta. His responsibilities include coordinating and instructing multiple international field programs annually offered in Europe, East Asia, and Southeast Asia, as well as instructing human, urban, and physical geography courses on campus. His current research focuses on natural disaster aid, recovery, and redevelopment, as well as international and post-secondary field program scholarship and practice.